Name: _____

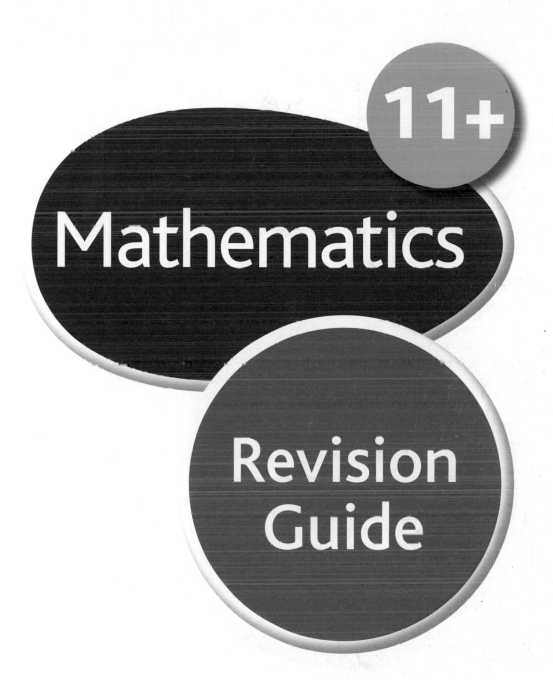

Mathematics

11+

Revision Guide

Louise Martine

GALORE PARK

AN HACHETTE UK COMPANY

Every effort has been made to trace all copyright holders, but if any have been inadvertently overlooked the publishers will be pleased to make the necessary arrangements at the first opportunity.

Although every effort has been made to ensure that website addresses are correct at time of going to press, Galore Park cannot be held responsible for the content of any website mentioned in this book. It is sometimes possible to find a relocated web page by typing in the address of the home page for a website in the URL window of your browser.

Hachette UK's policy is to use papers that are natural, renewable and recyclable products and made from wood grown in sustainable forests. The logging and manufacturing processes are expected to conform to the environmental regulations of the country of origin.

Orders: please contact Bookpoint Ltd, 130 Milton Park, Abingdon, Oxon OX14 4SB. Email: education@bookpoint. co.uk Telephone: +44 (0)1235 827720. Fax: (44) 01235 400454. Lines are open 9.00a.m.–5.00p.m., Monday to Saturday, with a 24-hour message answering service. Visit our website at www.galorepark.co.uk for details of other revision guides for Common Entrance, examination papers and Galore Park publications.

ISBN: 978 1 471849 21 3

Text copyright © Louise Martine 2016

First published in 2016 by

Galore Park Publishing Ltd,

An Hachette UK Company
Carmelite House
50 Victoria Embankment
London EC4Y 0DZ

www.galorepark.co.uk

Impression number   10 9 8 7 6 5 4 3 2

Year   2020 2019 2018 2017

Illustrations by Integra Software Services, Ltd.

Typeset in India

Printed in Spain

A catalogue record for this title is available from the British Library.

Contents and progress record

Use this page to plot your revision. Colour in the boxes when you feel confident with the skill and note your score and time for each test in the boxes.

/ 25   :

4 Measures, shape and space

| | Revised | Score | Time |

4

5   Algebra

Revised	**Score**	**Time**

6   Handling data

Revised	**Score**	**Time**

How to use this book

Introduction

This book has been written to help you revise the skills you have learned in Mathematics.

Use the book in the best way to help you revise. Work through the pages with a parent or on your own, then try the questions and talk about them afterwards. Each topic is presented as a two page spread which can be studied in half an hour. You are more likely to remember the skills and enjoy revising them in short bursts rather than spending a whole afternoon when you are tired. Try setting time aside after school two or three days a week. You may be surprised how quickly you will be able to revise the skills you have learned.

Pre-Tests and the 11+ entrance exams

The Galore Park 11+ series is designed for Pre-Tests and 11+ entrance exams for admission into independent and grammar schools. There are now several different kinds of mathematical test and it is likely that, if you are applying to more than one school, you will encounter more than one type of test. These include:

- Pre-Tests delivered on-screen
- 11+ entrance exams in different formats from GL, CEM and ISEB
- 11+ entrance exams created specifically for particular independent schools.

Calculators will not be allowed in nearly all 11+ tests, so mental arithmetic skills are very important. In some tests, mental arithmetic is tested through an oral paper, read out to the children.

The format and types of question can differ from year to year, both within each test and across the range of papers from test providers. This means a wide range of question types can occur across each test. This book covers the main types of question now typically occurring and provides a little practice to increase your speed.

The learning ladders

These ladders appear throughout the book to chart your progress through the areas of mathematics covered in each chapter. Your mathematical skills and ability to solve problems will develop as you step up the ladder, bringing together all of your knowledge to solve the most challenging questions by the time you reach the top.

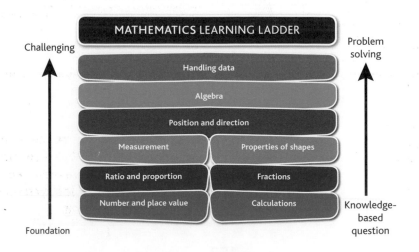

Working through the book

The contents and progress record helps you to keep track of your progress. When you have finished one of the spreads or tests complete the progress record by:

- colouring in or ticking the 'Revised' box on the planner when you are confident you have mastered the skill
- adding in your test scores and time to keep track of how you are getting on and to remind you of the areas in which you may need more practice.

Chapters link together related skills.

Chapter introductions explain how you can boost your mathematical skills.

Learning pages in Chapters 1–6 introduce related skills.

Train

A few straightforward questions will introduce you to using the skills. Write all your answers to the questions in an exercise book.

Key words are listed at the end of the learning section. It is a good idea to make sure you know and understand the definition of each of these words.

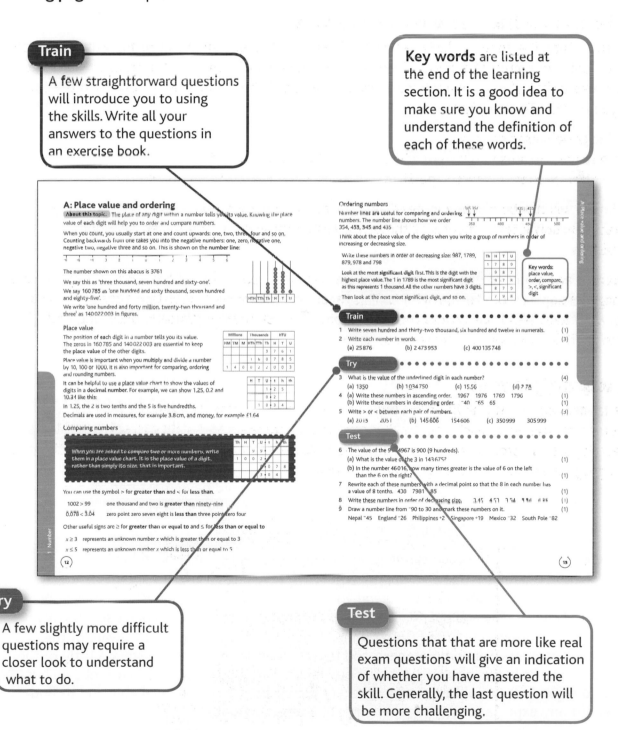

Try

A few slightly more difficult questions may require a closer look to understand what to do.

Test

Questions that that are more like real exam questions will give an indication of whether you have mastered the skill. Generally, the last question will be more challenging.

- **End of chapter tests** give you a chance to try out the skills you have just revised. Always time yourself to build up your speed. The test times given are representative of the average time given in an 11+ test.
  - Go through the test again with a friend or parent and talk about the questions you found tricky.
  - Later on, have a second attempt using the 'test' timing to get an idea of how quick the test might be under exam conditions.
- **11+ sample test:** The test at the end of the book is based on a test that could be set by an independent school. You may find some of the questions hard, but don't worry. This is a training test to build up your skills. Agree with your parents on a good time to take the test and set a timer going. Prepare for the test as if you are actually going to sit your 11+ (see 'Test day tips' below).
  - Complete the test with a timer, in a quiet room, noting down how long it takes you, writing your answers in pencil.
  - Mark the text using the answers at the back of the book.
  - Go through the test again with a friend or parent and talk about the difficult questions.
  - Have another go at the questions you found difficult and read the answers carefully to find out what to look for next time.
  - If you didn't finish the test in the given time, have another attempt before moving on to more practice tests in the Galore Park *Practice Papers* books.
- **Answers** to all the tests in the book can be found in the cut-out section beginning on page 173: Try not to look at the answers until you have attempted the questions yourself.

Test day tips

Take time to prepare yourself the day before you go for the test: remember to take sharpened pencils, an eraser, protractor, ruler, compasses and a calculator if it is allowed. A watch is very important so you can keep track of the time. Take a bottle of water in with you, if this is allowed, as this helps to keep your brain fresh and improves your concentration levels … and don't forget to have breakfast before you go!

For parents

This book has been written to help you and your child prepare for both Pre-Tests and 11+ entrance exams. Explanations at the beginning of each chapter include:

- tips on important mathematical skills that your child should have mastered
- advice on how additional work can have an impact on success
- activities and games for using the skills in enjoyable ways.

The teaching content in each section is designed so that it can be tackled in simple steps. All the explanations and examples tie into the teaching given. This approach has been followed in order to support you and your child in reviewing the questions they may have found challenging and provide ideas for points to look out for when practising further questions. Syllabuses are revised from time to time so it is important to be aware of the latest arrangements. For completeness, and for the interests of high achievers, a little of the material covered in this book may be just outside the requirements of the current syllabus. It is also a good idea to obtain copies of recent past papers if any are available and to find out about any expectations or requirements of a target school.

Continue your learning journey

When you've completed this *Revision Guide*, you can carry on your learning and practising with the following resources.

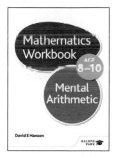 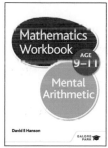  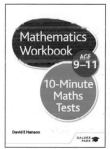  

The *Workbooks* will further develop your skills with plenty of practice questions:

● *Mental Arithmetic* tests will challenge you to think quickly and accurately without the aid of a calculator or the option to write down your working.
● *10-Minute Maths Tests* are a mix of topic-specific tests and mixed-topic tests, allowing you to identify which topics you are weakest in and focus on improving in those areas.

Practice Papers 1 contains four training papers, four short-format papers and one longer length paper with answers to improve your accuracy, speed and ability to deal with a wide range of questions.

Practice Papers 2 contains a further four model papers and answers to improve your accuracy, speed and ability to deal with a wide range of questions under pressure.

1 Number

Introduction

We are surrounded by numbers: they form an important part of everyday life. How could you find your favourite television channel, tell the time or use your calculator without numbers?

Mankind has questioned the nature of both things and people throughout its history. It seems likely that prehistoric humans could distinguish between, for example, one and two mammoths. Notches on bones found in Africa provide evidence of early man using and recording numbers. As societies developed, so did the use of symbols and words to represent numbers and patterns.

The Roman system for numerals was used throughout Europe for many years. The symbols for the first 10 counting numbers, the tens and the hundreds, shown below, are used to form all the numbers between 1 and 1000. Try to write some of the numbers between 1 and 1000 using Roman numerals – it is a good exercise in mental arithmetic!

I	II	III	IV	V	VI	VII	VIII	IX	X
1	2	3	4	5	6	7	8	9	10

X	XX	XXX	XL	L	LX	LXX	LXXX	XC	C
10	20	30	40	50	60	70	80	90	100

C	CC	CCC	CD	D	DC	DCC	DCCC	CM	M
100	200	300	400	500	600	700	800	900	1000

In this chapter, you will revise some of the important properties of numbers. These are at the bottom of the ladder. Understanding them will help you as you work through the book and climb the ladder.

Learn the meaning of all the **key words** and aim to answer all the questions confidently before you move on to the next skill. As you work through the book, you will be reminded of all the interesting things you can do with numbers. Believe in your abilities!

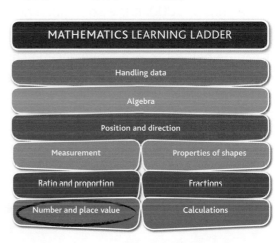

MATHEMATICS LEARNING LADDER

Handling data

Algebra

Position and direction

Measurement

Properties of shapes

Ratio and proportion

Fractions

Number and place value

Calculations

Problem solving

Don't be put off by questions that are set in the context of real life. The calculations and numbers you need are simply hidden within a story. Write down the key information and decide which calculation(s) you are being asked to do. Then work out the answers, writing down your working. You will practise solving problems in the '**test**' sections at the end of each skill.

Advice for parents

This chapter promotes an understanding of the basic properties of numbers. Working through this chapter slowly and carefully will help ensure that important skills such as rounding and estimating have been mastered. Estimating should be used to check that actual answers are sensible and no silly mistakes have been made in calculations. Further checking that all answers are written in the correct format is encouraged as this is another area where valuable marks are frequently lost. Finally, the importance of secure recall of **times tables** and learning the **rules of divisibility** cannot be overestimated.

A: Place value and ordering

About this topic: The place of any digit within a number tells you its value. Knowing the place value of each digit will help you to order and compare numbers.

When you count, you usually start at one and count upwards: one, two, three, four and so on. Counting backwards from one takes you into the negative numbers: one, zero, negative one, negative two, negative three and so on. This is shown on the **number line**:

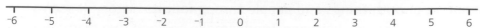

The number shown on this abacus is 3761

We say this as 'three thousand, seven hundred and sixty-one'.

We say 160 785 as 'one hundred and sixty thousand, seven hundred and eighty-five'.

We write 'one hundred and forty million, twenty-two thousand and three' as 140 022 003 in figures.

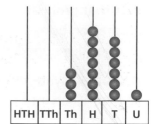

Place value

The position of each digit in a number tells you its value. The zeros in 160 785 and 140 022 003 are essential to keep the **place value** of the other digits.

Place value is important when you multiply and divide a number by 10, 100 or 1000. It is also important for comparing, ordering and rounding numbers.

Millions			Thousands			HTU		
HM	TM	M	HTh	TTh	Th	H	T	U
					3	7	6	1
			1	6	0	7	8	5
1	4	0	0	2	2	0	0	3

It can be helpful to use a place value chart to show the values of digits in a **decimal number**. For example, we can show 1.25, 0.2 and 10.34 like this:

H	T	U	t	h	th
		1	2	5	
		0	2		
	1	0	3	4	

In 1.25, the 2 is two tenths and the 5 is five hundredths.

Decimals are used in measures, for example 3.8 cm, and money, for example £1.64

Comparing numbers

When you are asked to compare two or more numbers, write them in a place value chart. It is the place value of a digit, rather than simply its size, that is important.

Th	H	T	U	t	h	th
		9	9			
1	0	0	2			
			0	0	7	8
			3	0	4	

You can use the symbol > for **greater than** and < for **less than**.

1002 > 99   one thousand and two is **greater than** ninety-nine

0.078 < 3.04   zero point zero seven eight is **less than** three point zero four

Other useful signs are ≥ for **greater than or equal to** and ≤ for **less than or equal to**

$x \geq 3$ represents an unknown number x which is greater than or equal to 3

$x \leq 5$ represents an unknown number x which is less than or equal to 5

Ordering numbers

Number lines are useful for comparing and **ordering** numbers. The number line shows how we order 354, 453, 345 and 435

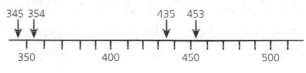

Think about the place value of the digits when you write a group of numbers in order of increasing or decreasing size.

Write these numbers in order of decreasing size: 987, 1789, 879, 978 and 798

Look at the most **significant digit** first. This is the digit with the highest place value. The 1 in 1789 is the most significant digit as this represents 1 thousand. All the other numbers have 3 digits.

Then look at the next most significant digit, and so on.

Th	H	T	U
1	7	8	9
	9	8	7
	9	7	8
	8	7	9
	7	9	8

Key words:
place value, order, compare, >, <, significant digit

Train

1   Write seven hundred and thirty-two thousand, six hundred and twelve in numerals. 732,6(1)12   (3)

2   Write each number in words.   (3)

(a) 25 876   (b) 2 473 953   (c) 400 135 748

twenty five thousand and eight hundred and seventy-six.

Try

3   What is the value of the underlined digit in each number?   (4)

(a) 135<u>0</u> *tens*   (b) 1<u>0</u>34 750 *hundred thousand*   (c) 15.5<u>6</u> *fifty*   (d) 2.7<u>8</u> *ones*

4   (a) Write these numbers in ascending order. 1967 1976 1769 1796   (1)
    (b) Write these numbers in descending order. ⁻40 ⁻65 65   (1)

5   Write > or < between each pair of numbers.   (3)

(a) 2015 < 2051   (b) 145 606 < 154 606   (c) 350 999 > 305 999

Test

6   The value of the 9 in 4967 is 900 (9 hundreds).

(a) What is the value of the 3 in 143 675? *thousand*   (1)

(b) In the number 46 016, how many times greater is the value of 6 on the left than the 6 on the right? *6 thousand*   (1)

7   Rewrite each of these numbers with a decimal point so that the 8 in each number has a value of 8 tenths.   438   7981 0.85   (1)

8   Write these numbers in order of decreasing size.   3.45   4.53   3.54   5.34   4.35   (1)

9   Draw a number line from ⁻90 to 30 and mark these numbers on it.   (1)

Nepal ⁻45   England ⁻26   Philippines +2   Singapore +19   Mexico ⁻32   South Pole ⁻82

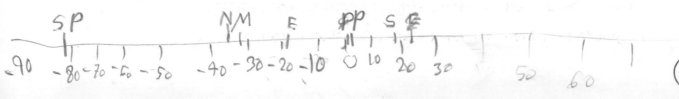

B: Rounding and estimating

About this topic: Rounding numbers is a very useful skill. You can use rounded values to estimate the results of a calculation and use this to check your actual answer is sensible.

Rounding integers

You can round numbers to the **nearest ten, hundred, fifty** and so on.

If the value is **below** the halfway mark, then you **round down**. This is 0, 1, 2, 3 or 4 if you are rounding a units digit to the nearest ten.

If the value is **at or above** the halfway mark, then you **round up**. This is 5, 6, 7, 8 or 9 if you are rounding a units digit to the nearest ten.

Look at these examples from everyday life.

If we are sharing 18 sweets between four children, they will receive four and a half sweets each.

In this case it is sensible to round down. Each child will receive four sweets and there will be two sweets left over.

If seven people are going to eat half a pizza each, then altogether they will eat three and a half pizzas.

In this case it is sensible to round up. They will need to buy four pizzas.

(i) Round 345 to the nearest ten.
Imagine a line between the tens and the units columns.

Th	H	T	U
	3	4	5
	3	5	0

The digit to the right of this line is 5, so you round up. Add 1 to the tens column (left of the vertical line) and replace each digit to the right with a zero. 345 rounded to the nearest ten is 350

(ii) Round 1549 to the nearest hundred.
Imagine a line between the hundreds and the tens columns.

Th	H	T	U
1	5	4	9
1	5	0	0

The number to the right of the line is 4, so replace each digit to the right with a zero. 1549 rounded to the nearest hundred is 1500

You can use the symbol ≈ which means 'approximately equal to' when you round a number. You can write the answer to the last example as 1549 ≈ 1500

Rounding decimals

The rules for rounding decimals are the same as for rounding integers. Look at these examples.

Rounding to the nearest integer (whole number)

Rounding 7.4 to the nearest integer gives 7

Rounding 7.49 to the nearest integer gives 7

Rounding 7.5 to the nearest integer gives 8

Rounding 7.6 to the nearest integer gives 8

> When you round decimals to the nearest whole number, do not simply replace the digits to the right of the decimal point with zeros. You must look at whether the value of the decimal is less than, or halfway or more, to the next whole number.

Rounding to a number of decimal places

2.444 is	**2.44**	written to 2 decimal places (2 d.p.)
	2.4	written to 1 d.p.
7.245 is	**7.25**	written to 2 d.p.
	7.2	written to 1 d.p.
6.993 is	**6.99**	written to 2 2 d.p.
	7.0	written to 1 d.p.

Rounding to a number of significant figures

The same basic rules apply when rounding any number (integer or decimal) to a number of significant figures.

A significant figure is a digit that is part of a number. It tells you how many units, tens, hundreds, and so on there are in that number.

Consider 402 355

4 is the first significant figure. It shows the number has 4 hundred thousands.

0 is the second significant figure. It shows the number has no ten thousands.

2 is the third significant figure. It shows the number has two thousands.

When you are rounding to a number of significant figures, you must first identify which digit you are considering.

Look at the examples.

6594 is	**659**0	written to 3 significant figures (s.f.)
	6600	written to 2 s.f.
	7000	written to 1 s.f.

> Remember to put zeros in where required. The zeros in the example above are not significant, but they are needed to keep the place values of the other digits.

15

Zeros on the right are *sometimes* significant.

39.57 is   **39.6**   to 3 s.f.

          40   to 2 s.f. (the zero is significant)

          40   to 1 s.f. (the zero is not significant)

Zeros in the middle are *always* significant.

0.6032 is   0.60   to 2 s.f.

          0.6   to 1 s.f.

304.03 is   **304.0**   to 4 s.f.

          300   to 2 s.f.

Zeros on the left with no non-zero digits before them are *never* significant.

0.6032 is   0.**603**   to 3 s.f.

Estimating numbers and calculations

You do not always need to give exact values when you are talking about numbers.

You may not always know an exact number or measure. You can estimate a number of **objects**, for example the number of people watching a rugby match, the number of sweets in a jar or the number of 1 penny coins you could fit onto a sheet of A4 paper.

When estimating a large number of things, it is important to be able to explain your answer rather than just taking a wild guess.

Measurements can be estimated. For example, you might estimate the length of a piece of string. A length, an area, a volume, a mass, an angle or indeed any other measurement can be estimated.

> An estimate is only approximate. It is a rounded number.

Remember that all measurements are approximate. The length of a pencil could be measured as 9.7 cm, but it could really be 9.69 cm or 9.71 cm.

Rounding numbers also allows you to make an estimate of the result of a **calculation**.

If Emily is thinking about buying a fancy hat costing £4.95 for each of the 11 guests at her party, she could estimate by rounding the numbers to £5 and 10

This would give an estimated cost of £50

The cost would really be £54.45 but the estimate is all Emily needs so that she knows roughly what the total cost of the fancy hats would be.

Estimate the result of 395×41

First round the values. $395 \approx 400$ (round up to the nearest hundred) and
                          $41 \approx 40$ (round down to the nearest ten)

Then do the calculation mentally. $400 \times 40 = 16\,000$

Computers cost £375 each. The school buys a new computer for each of its 11 classrooms. Estimate the cost of the computers.

£375 is £400 to the nearest £100

11 is 10 to the nearest 10

An estimate of the cost is £400 × 10 = £4000

> **Key words:** integer, decimal place, significant figure, estimate, rounding

Train

1 Round each number to the nearest hundred.   (3)
   (a) 671 _700_   **(b)** 1425 _1 400_   **(c)** 452 750 _452 800_

2 Round each number to the nearest twenty.   (3)
   (a) 1558 _1 600_   **(b)** 27 455   **(c)** 413 124

3 Round each decimal to the nearest integer.   (3)
   (a) 8.34   **(b)** 9.53   **(c)** 5.79

Try

4 Round these numbers to:
   (i) 1 d.p.   **(ii)** 2 d.p.   **(iii)** the nearest whole number   **(iv)** 2 s.f.   (16)

   (a) 7.245   **(b)** 3.482   **(c)** 10.568   **(d)** 43.479

Test

5 Estimate the answer to each calculation.
   (a) 24 713 + 54 357   (Round each number to the nearest thousand)   (2)
   (b) 733 126 – 454 653   (Round each number to the nearest ten thousand)   (2)
   (c) 596 × 31   (2)
   (d) 905 ÷ 9.7   (2)

6 The heights of three mountains are:

   Mount Everest   8848 m   K2   8611 m   Kanchenjunga   8586 m

   (a) Estimate the difference in height between Mount Everest and K2 by first rounding both heights to the nearest one hundred metres.   (2)

   (b) Estimate the difference in height between Kanchenjunga and K2 by first rounding both heights to the nearest ten metres.   (2)

   (c) Estimate the average height of the three mountains by first rounding each height to the nearest one hundred metres.   (2)

C: Index numbers, roots, factors and multiples

About this topic: In this section, you will look at index numbers, square roots and cube roots. You will also revise how to find factors and multiples of numbers, including rules for divisibility.

Index numbers

The index of a number tells you how many 'lots' of that number are multiplied together. The **index number** is written as a small raised number.

4^2 is 4×4 and called four **squared** (index number 2)

5^3 is $5 \times 5 \times 5$ and called five **cubed** (index number 3)

We say index numbers greater than 3 as 'to the power of ...'

6^4 is $6 \times 6 \times 6 \times 6$ and called six **to the power of** 4

Writing large numbers in index form can sometimes make them easier to deal with.

$1\,000\,000 = 10 \times 10 \times 10 \times 10 \times 10 \times 10 = 10^6$ One million can be written as 10^6

Square numbers appear in the diagonal of the multiplication table. They are formed by multiplying a number by itself, $n \times n = n^2$. We say that n^2 is 'the square of n' or 'n squared'.

$1^2 = 1$ $2^2 = 4$ $3^2 = 9$ $4^2 = 16$ ...

Roots

The **square root** of an integer is the number which, when multiplied by itself, will give that integer.

The square root of 4, written as $\sqrt{4}$, is 2 because $2 \times 2 = 4$ If $n^2 = 16$, then $n \times n = 16$ and $\sqrt{16} = 4$ so $n = 4$

Other powers also have associated roots.

The **cube root** of 8 is 2 because $2 \times 2 \times 2 = 8$ This is written as $\sqrt[3]{8} = 2$

Factors and multiples

> Write out a 12 × 12 multiplication table square. Study it until you are confident that you understand all the patterns within it.

Multiples are formed by multiplying an integer (a factor) by another integer.

12 is a multiple of both 3 and 4 because $3 \times 4 = 12$

56 is a multiple of 7 and 8 and, since it is in both the 7 and 8 times tables, we can say 56 is a **common multiple** of 7 and 8

A number that divides exactly into another number is called a **factor** of that number. Since $3 \times 4 = 12$, 3 and 4 are both **factors** of 12

3 and 4 form a **factor pair** of 12 because they multiply together to make 12

A number can have one or more pairs of factors.   12 has three factor pairs. 2×6 3×4 12×1

You can draw a **factor rainbow** to show all the factor pairs of a number.

56 has four factor pairs.

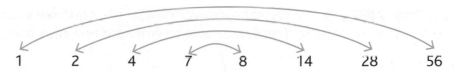

1 2 4 7 8 14 28 56

Rules of divisibility

Learning the rules of divisibility will help you to find the factors of a number.

A number is divisible by...	Example
2 if the last digit is even (0, 2, 4, 6, 8,)	$12 \div 2 = 6$
3 if the digit sum is a multiple of 3	Digit sum of 15 is $1 + 5 = 6$ and 6 can be divided exactly by 3
4 if the last two digits can be divided by 4	6416 divisible by 4 because $16 \div 4 = 4$
5 if the last digit is 0 or 5	1630 is divisible by 5 because the last digit is 0
6 if it is an even number and the digit sum is a multiple of 6	Digit sum of 48 is $4 + 8 = 12$ and 12 can be divided exactly by 6
9 if the digit sum is a multiple of 9	Digit sum of 918 is $9 + 1 + 8 = 18$ and 18 can be divided exactly by 9
10 if the last digit is 0	$780 \div 10 = 78$

Key words: square, cube, 'to the power of', square root, cube root, factor, common factor, multiple

Train

1 Work out the value of: (a) 2^6 (b) $\sqrt{25}$ (c) $\sqrt[3]{27}$ (3)

2 Write down all the factors pairs of: (a) 24   (b) 36   (c) 45   (3)

Try

3 Copy each pair of numbers and write the correct symbol (<, > or =) between them.   (4)
   (a) 3^6 ... 6^2 (b) 80 ... 9^2 (c) $\sqrt{25}$ ... 5   (d) 8 ... $\sqrt[3]{64}$

4 (a) Draw a factor rainbow for 68   (b) Write down all the factor pairs.   (2)

5 What is the lowest number that has both 3 and 9 as factors?   (1)

Test

6 If $a^b = 64$ what are the values of a and b?   (1)

7 Here are some number cards.

   | 4 | 5 | 11 | 12 | 15 | 24 | 27 | 33 | 35 |

   Choose from the numbers above: (7)
   (a) a multiple of 7
   (b) a factor of 10
   (c) a square number
   (d) the number that can be written as $2 \times 2 \times 2 \times 3$
   (e) the number that is a common multiple of 3 and 5
   (f) two numbers that have a product of 48
   (g) a cube number

D: Prime numbers, prime factors, HCF and LCM

About this topic: In this section, we will look at some other important properties of numbers. It is often helpful in mathematics to be able to write an integer as the product of its prime factors.

Prime numbers

A prime number has only two factors, 1 and itself. 1 is not a prime number, because it has only one factor. 2 is the only even prime number.

The prime numbers up to 100 are:

2 3 5 7 13 17 19 23 29 31 37 41 43 47 53 59 61 67 71 73 79 83 89 97

Prime factors

The **prime factors** of an integer are the factors of that integer that are prime numbers. Every integer can be written as the product of its prime factors.

The factors of 24 are 1, 2, 3, 4, 6, 8, 12 and 24

The prime factors of 24 are 2 and 3

We can write 24 as $2^3 \times 3$

We can use factor trees or ladder division to find the prime factors of a number. Use ladder division to find the prime factors of larger numbers.

Factor tree

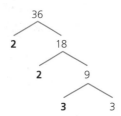

The prime factors of 36 are 2 and 3

As a product of its prime factors:

$36 = 2 \times 2 \times 3 \times 3$

Ladder division

Start by dividing by 2

When you can no longer divide by 2, move on to the next smallest prime number. If this does not divide exactly, move on to the next smallest prime until you get to 1

The prime factors of 320 are 2 and 5

As a product of its prime factors

$320 = 2 \times 2 \times 2 \times 2 \times 2 \times 2 \times 5$
$\qquad = 2^6 \times 5$ (using index numbers)

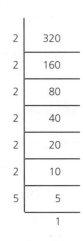

Highest common factor (HCF)

The **highest common factor (HCF)** of two numbers is the largest number that is a factor of both of the numbers. To find the HCF of two numbers, write down the factors of each number and then compare the two lists. A Venn diagram can be useful in showing common factors.

> You can use factor rainbows to show all the factor pairs and therefore all the factors of a number.

What is the highest common factor of 24 and 40?

The factors of 24 are 1, 2, 3, 4, 6, 8, 12, 24

The factors of 40 are 1, 2, 4, 5, 8, 10, 20, 40

1, 2, 4 and 8 appear as factors in both numbers. 8 is the largest number in both lists, so 8 is the highest common factor of 24 and 40

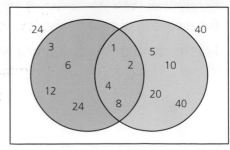

Lowest common multiple (LCM)

The **lowest common multiple** of two numbers is the smallest number that is a multiple of both of the numbers.

You can always find a **common multiple** by multiplying both numbers together, but it is not always the LCM. In the example above, $10 \times 15 = 150$ which is not the LCM.

What is the LCM of 10 and 15?

The multiples of 10 are 10, 20, 30, 40, 50, ...

The multiples of 15 are 15, 30, 45, 60, ...

The LCM of 10 and 15 is 30

> **Key words:** prime numbers, prime factors, factor trees, ladder division, highest common factor, lowest common multiple

Train

1   How many of these are prime numbers? (1)

   1   2   3   19   27   49   63   79   91   111

   A: 3   **B:** 4   **C:** 5   **D:** 6   **E:** 7

2   Write each number as the product of its prime factors in index form. (3)

   (a) 24   **(b)** 26   **(c)** 27

3   **(a)** What is the highest common factor of 12 and 20? (1)

   (b) What is the lowest common multiple of 12 and 18? (1)

Try

4   Draw a 10 × 10 number square with numbers from 101 to 200

   Use your number square to work out all the prime numbers between 101 and 200

   Hint: If you cross out the multiples of 2, 3, 5, 7, 11 and 13, you will be left with the prime numbers. (1)

5   Write each number as the product of its prime factors. Write your answers in index form. (3)

   (a) 225   **(b)** 234   **(c)** 480

6   **(a)** What is the highest common factor of 100 and 150? (2)

   (b) What is the lowest common multiple of 5, 9 and 15? (3)

Test

7   Write 680 as a product of its prime factors in index form. (1)

8   Lighthouse A flashes every 8 seconds. Lighthouse B flashes every 12 seconds. If they start flashing at the same time, how long will it be until they next flash together? (1)

9   I have a pile of 5p coins and a pile of 20p coins. Both piles have the same value. What is the least amount of money I could have in total? (1)

Test 1

1 When these numbers are arranged in ascending order, which
   one is in the middle? (1)

   A: 3405   **B:** 0.345   **C:** 340   **D:** 3.45   **E:** 3.045

2 How many of these numbers are multiples of 3? (1)

   18   45   54   65   7   8   102   123   666   900   10 101

   A: 5   **B:** 6   **C:** 7   **D:** 8   **E:** 9

3 (a) How many times more than 456 is 45 600? (1)

   (b) How many times more than 7.14 is 7140? (1)

   (c) How many times more than 0.05 is 500? (1)

   (d) How many times more than 7.4 is 7400? (1)

   (e) How many times more than 1.09 is 109 000? (1)

4 Write down the value of each expression. (2)

   (a) $4^3 + 5^2$ (b) $\sqrt[3]{1000} - \sqrt{64}$

5 Estimate by first rounding the values as indicated in the brackets.

   (a) 1 478 456 + 3 405 124   (nearest ten thousand) (1)

   (b) 5 567 090 − 2 545 995   (nearest thousand) (1)

6 (a) What are the prime factors of 18 and 32? (1)
   (b) What is the highest common factor of 18 and 32? (1)

7 What is the largest number that will divide exactly into 48, 75 and 120? (2)

8 What is the lowest common multiple of 24 and 56? (2)

9 (a) Some pupils grew beans as part of a science experiment.
       The table shows the heights of their plants after three weeks.

Name	Height of plant (cm)
Charles	35.66
Rosanna	43.25
Peter	45.04
Cecelia	29.89

   (i) List the heights in ascending order. (1)
   (ii) Round each height to the nearest integer. (2)
   (iii) Round each height to 1 d.p. (2)
   (iv) Write each height to 1 s.f. (2)

   (b) The beans grow taller and need to be tied up with string. The pupils have two pieces of
       string: one is 24 cm long and the other 18 cm long. They want to cut these pieces into
       equal lengths. What is the longest possible length they can cut? (2)

   (c) Rosanna and Charles have set up automated systems to water their plants. They come on
       together at 9.00 a.m. Rosanna's system releases water every 2 hours and Charles' system
       releases water every 3 hours. At what time will they next release water together? (2)

**Record your score and time here
and at the start of the book.**

Score [   ] / 28   Time [ ] : [ ]

② Calculations

Introduction

Calculations are everywhere. How much money do you need to buy three batteries? How many days until the end of term? How do you divide a pizza equally amongst four friends?

There are **four basic operations**.

- **Addition:** the words associated with addition are total, more, together, added to, increased, greater, up.
- **Subtraction:** the words associated with subtraction are less, take away, fewer, lower, down, from.
- **Multiplication:** the words associated with multiplication are product, times, lots of, groups of, rows of.
- **Division:** the words associated with division are goes in to, shared, split equally.

Addition and subtraction facts come in groups of four. You can use one fact in a group to work out the other three.

$$3 + 7 = 10 \qquad 10 - 7 = 3$$
$$7 + 3 = 10 \qquad 10 - 3 = 7$$

Multiplication and division facts also come in groups of four. Again, you can use one fact in a group to work out the other three.

$$3 \times 7 = 21 \qquad 21 \div 7 = 3$$
$$7 \times 3 = 21 \qquad 21 \div 3 = 7$$

These **five rules** can help you **avoid errors** with calculations.

1 Make a mental estimate of your answer so you know roughly what to expect.
2 Set out your calculations clearly and neatly.
3 Take care with place value.
4 Check that your answer is sensible.
5 Write your answer using the correct units where needed.

In word problems, identify the calculation first and then work it out. Pay attention to signs and symbols such as index numbers, negative signs and brackets.

> *Doing a different calculation to check your answer is useful if you have time.*

In this chapter, you will be working through the skills on the second half of the first step of the ladder.

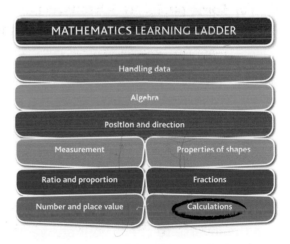

MATHEMATICS LEARNING LADDER

Handling data

Algebra

Position and direction

Measurement — Properties of shapes

Ratio and proportion — Fractions

Number and place value — Calculations

Advice for parents

This chapter starts by covering mental strategies. Being able to use a range of these strategies enables easier manipulation and computation of calculations. There are no rules about which method to use, so not everyone will approach a mental calculation in the same way. The important thing is that the answer is correct!

Mastering calculations in mathematics is easier if addition and subtraction facts and times tables are known. Writing multiplication or division facts on coloured cards can support learning if they are displayed where they can be seen often.

A: Mental strategies

About this topic: Revise a range of mental strategies to calculate without a calculator or pen and paper. There is no one correct method.

Addition

When adding numbers together, the order does not matter.
(The operation is **commutative**.)

$$6 + 3 = 9 \quad \text{and} \quad 3 + 6 = 9$$

Subtraction

When subtracting, the order of numbers does matter.
(The operation is *not* commutative.)

$$6 - 3 = 3 \quad \text{BUT} \quad 3 - 6 = {}^-3$$

Make sure you distinguish between a **negative** number (a single number below zero, for example $^-3$) and the **minus** sign (the operation of subtraction, for example $7 - 4$)

Mental strategies for addition and subtraction

- Picture the calculation in columns and deal with each column in turn from the right.
  For example: units then tens then hundreds.
- **Partition** a number, for example into tens and units.

> Always estimate before you calculate. Use your estimate to check whether or not your exact answer is sensible.

$56 + 23$ partition $23 = (20 + 3)$

So, $56 + 23 = 56 + (20 + 3) = 76 + 3 = 79$

- Imagine a **number line** and break the calculation into manageable stages.

$78 + 28 = 78 + 2 + 26 = 106$

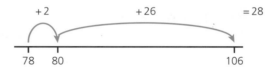

- Subtract by **counting on** or **counting back**.

$806 - 297 = 509$

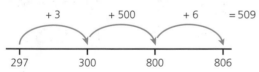

$81 - 27 = 54$

Use the inverse to check your answer when you subtract. $81 - 27 = 54$,
so check: $54 + 27 = 81$ ✔

Multiplication

When multiplying numbers together, the order does not matter.

$$^-6 \times {}^-2 = 12 \quad \text{and} \quad {}^-2 \times {}^-6 = 12$$

> Make sure you know your times tables really well.

Division

The order of numbers in division calculations *does* matter. $^-6 \div {}^-2 = 3$ BUT $^-2 \div {}^-6 = \frac{1}{3}$

When a division is not exact, there is a remainder. $12 \div 5 = 2$ remainder 2

Mental strategies for multiplication and division

- Use **multiplication and division facts.** $3 \times 7 = 21$ $7 \times 3 = 21$ $21 \div 7 = 3$ $21 \div 3 = 7$

- Make use of **known facts.** What is $1.44 \div 1.2$? $144 \div 12 = 12$, so $1.44 \div 1.2 = 1.2$

- Use **doubling** and **halving**. It is sometimes better to double or halve several times, than to attempt a more complicated calculation.

  Double 15, double again and double again, then subtract the original number to multiply by 7

  $15 \times 7 \rightarrow 30\ (15 \times 2)$, then 60, then 120, then 105 $(120 - 15)$

  £34.60 ÷ 4 → £34.60 ÷ 2 ÷ 2   £34.60 ÷ 2 = £17.30   £17.30 ÷ 2 = £8.65   £34.60 ÷ 4 = £8.65

- Think of 5 as half of 10

  $575 \div 5 \rightarrow 575 \times 2 \div 10 = 1150 \div 10 = 115$

- Use **partitioning** to multiply by a larger number.

  $13 \times 7 = (10 \times 7) + (3 \times 7) = 70 + 21 = 91$

> Use division to answer questions that ask you to find '$\frac{1}{2}$ of', a '$\frac{1}{3}$ of' and so on. Finding '$\frac{1}{2}$ of' an amount is the same as dividing by 2

Key words: partition, negative number, minus sign

Train

1 Revise the four basic operations by calculating the answers to the following. (10)

(a) 12 + 6	(c) $^-$12 + 6	(e) $^-$12 − 6	(g) 6 − $^-$12	(i) $^-$12 × $^-$6
(b) 6 + 12	(d) $^-$12 + $^-$6	(f) $^-$12 − $^-$6	(h) $^-$6 − $^-$12	(j) $^-$6 ÷ 12

Try

2 Calculate these answers mentally. (8)

(a) 105 + 44	(c) 146 − 34	(e) 2 × 16	(g) 140 ÷ 4
(b) 767 + 133	(d) 338 − 39	(f) 2 × 37	(h) 425 ÷ 5

Test

3 There were 48 people on bus. When the bus reached Oxford, 37 people got off and 23 people got on. How many people were then on the bus? (2)

4 $48 \div 3 = 16$ Use this fact to answer these questions. (4)

(a) 480 : 3	(b) 48 ÷ 30	(c) 4.8 ÷ 3	(d) 4.8 ÷ 6

B: Negative numbers

About this topic: Revise **negative** numbers in more detail before tackling more complex number calculations.

> Remember, negative 2 is a number below zero. Minus 2 is an operation linking two numbers. Minus 2 only has meaning when we say what we are taking 2 away from, for example 8 – 2

You will see negative numbers in the context of measurement, for example as temperatures below zero and depths below sea level, as well as in number calculations.

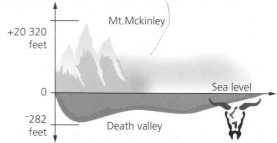

Number lines

In mathematics, we can show negative numbers on a number line. The position of ⁻4 is shown.

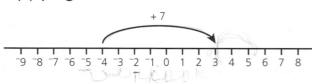

We can use a number line to:

● **add a positive integer to a negative integer**

⁻4 + 7 = 3

⁻9 + 6 = ⁻3

● **subtract a positive integer across zero**   ● **subtract a positive integer from a negative integer**

4 – 6 = ⁻2

⁻3 – 6 = ⁻9

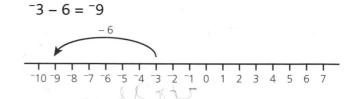

● **add a negative integer**
  (Adding a negative number is the same as subtracting the 'opposite' positive number.)

⁻5 + (⁻2) = ⁻7

3 + (⁻6) = −3

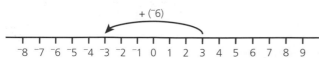

● **subtract a negative integer**
(Subtracting a negative number is the same as adding the 'opposite' positive number.)

$3 - (^-6) = 9$ $^-6 - (^-4) = ^-2$

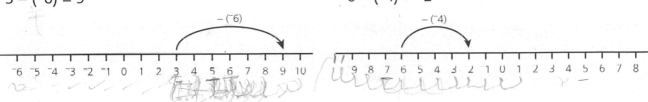

Number patterns

To complete number patterns, work out the difference between two adjacent numbers. Then add or subtract the step to find any missing terms.

Find the next two terms in $^-6, ^-3, 0, 3, ..., ...$

The difference is $^-3 - (^-6) = ^-3 + 6 = 3$ So add 3 each time.

The next two terms are $3 + 3 = 6$ and $6 + 3 = 9$

Follow the same steps for number patterns with decimal numbers (non-integers).

Find the next two terms in $^-2.7, ^-1.7, ^-0.7, ..., ...$

The difference is $^-0.7 - (^-1.7) = ^-0.7 + 1.7 = 1$ So add 1 each time.

The next two terms are $^-0.7 + 1 = 0.3$ and $0.3 + 1 = 1.3$

> **Key words:**
> negative, minus, positive

Train

1   Calculate the following. Use a number line if it helps.   (6)
   (a) $^-1 + 3$ **(b)** $^-5 + 1$ **(c)** $^-2 + 2$ **(d)** $6 - 8$ **(e)** $^-1 - 5$ **(f)** $^-3 - 10$

2   Calculate the following. Use a number line if it helps.   (6)
   (a) $5 + (^-2)$ **(b)** $3 + (^-4)$ **(c)** $^-7 + (^-3)$ **(d)** $5 - (^-5)$ **(e)** $6 - (^-4)$ **(f)** $^-3 - (^-5)$

Try

3   Calculate the following.   (6)
   (a) $^-11 + 12$ **(b)** $^-15 + 5$ **(c)** $^-7 + 14$ **(d)** $14 - 28$ **(e)** $25 - 35$ **(f)** $^-30 - 27$

4   Calculate the following.   (6)
   (a) $34 + (^-14)$ **(b)** $17 + (^-20)$ **(c)** $^-12 + (^-13)$ **(d)** $16 - (^-4)$ **(e)** $^-15 - (^-16)$

Test

5   Calculate the following.   (3)
   (a) $^-10 - (^-15)$ **(b)** $^-1 + (^-2) + (^-1)$ **(c)** $24 + (^-20) + (^-15)$

6   Work out the missing terms in each pattern.   (2)
   (a) $^-10, ^-2, 6, 14, ..., ...$ **(b)** $..., ..., 0. 0.6, 1.2, 1.8$

7   Average monthly temperatures in Antarctica: January $6\,°F$, May $^-23\,°F$ and July $^-29\,°F$.
   (a) What is the difference between the average temperatures in May and July?   (1)
   (b) What is the difference between the average temperatures in January and May?   (1)

C: Order of operations

About this topic: When there are several operations within a calculation, the order in which you do them is important.

Without rules, you could read $12 - 5 \times 4$ as $(12 - 5) \times 4 = 7 \times 4 = 28$ or $12 - (5 \times 4) = {}^-8$

The rule for dealing with calculations with **mixed operations** (a mix of addition, subtraction, multiplication and division) is that you do the calculation in the brackets first, if there are brackets, then multiply or divide before you add or subtract.

The mnemonic **BIDMAS** will help you remember this rule.

> B   Brackets
>
> I   Indices
>
> D   Divide
>
> M   Multiply
>
> A   Add
>
> S   Subtract

● Do the part(s) of the calculation inside the **brackets** first.

With brackets: $4 \times (8 - 6) = 4 \times 2 = 8$

Without brackets: $4 \times 8 - 6 = 32 - 6 = 26$

● Deal with any **indices** next.

$2^2 \times 5 = 4 \times 5 = 20$ Deal with the indices first.

● Then **division and multiplication** before **addition and subtraction**.

$4^2 + 5 \times (12 - 2) \div 10 - 2^3$

$$= 4^2 + 5 \times 10 \div 10 - 2^3 \qquad \text{Deal with the Brackets first}$$
$$= 16 + 5 \times 10 \div 10 - 8 \qquad \text{then the Indices}$$
$$= 16 + 5 - 8 \qquad \text{then Division and Multiplication}$$
$$= 13 \qquad \text{and finally Addition and Subtraction}$$

Remember with **problem solving**, you need to identify the calculation hidden in the sentence(s) first.

Pupils recorded how many pieces of fruit they had eaten over the weekend.
60 ate 1 piece of fruit, 15 ate 2 pieces of fruit and 6 ate 3 pieces of fruit.
How many pieces of fruit were eaten in total over the weekend?

So, we have been told:

60 pupils ate 1 piece of fruit $60 \times 1 = 60$ pieces of fruit

15 pupils ate 2 pieces of fruit $15 \times 2 = 30$ pieces of fruit

6 pupils ate 3 pieces of fruit $6 \times 3 = 18$ pieces of fruit

Now add the individual totals to find out how many pieces of fruit were eaten: $60 + 30 + 18 = 108$

Answer: A total of 108 pieces of fruit were eaten over the weekend.

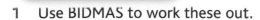

If the question is a sentence, give your answer as a short phrase or sentence.

> **Key words:** BIDMAS, brackets, indices

Train

1 Use BIDMAS to work these out. (8)

(a) $8 - (4 - 2)$ (c) $(9 \times 3) - (3 \times 7)$ (e) $90 \div 5 \times 2$ (g) $140 + 7 \div 7$

(b) $10 + (3 \times 6)$ (d) $9 + 7 - 3$ (f) $20 + 3 \times 2$ (h) $72 \div 6 - 120 \div 12$

Try

2 Calculate these. (4)

(a) $25 + (14 \times 2) \times 5$ (c) $84 \div 4 - 55 \div 5$

(b) $3^3 - (3 \times 4)$ (d) $5^2 + 6 \times (5 - 2) \div 3 - 2^3$

3 The art department ordered 15 pots of red paint, 10 pots each of blue paint and of yellow paint and 16 pots each of white paint and black paint.

How many pots of paint were ordered in total? (5)

Test

4 Which of these has the largest result? (5)

A: 6×9 B: 7×8 C: $110 \div 2$ D: $280 \div 5$ E: 29×2

5 Each pupil in a class of 21 was given 12 different poems to read over the summer holidays.

(a) Estimate the number of poems that were given out. (1)

(b) Work out the exact number of poems that were given out. (1)

6 The sum of two numbers is 25 and their product is 150

What are the numbers?

> The product is the result of multiplying numbers.

(2)

7 At the school summer fair:
the cake stall sold 25 cakes for £4 each
the plant stall sold 65 strawberry plants for £2 each
the jewellery stall sold 20 necklaces for £6 each.
How much money did these three stalls take in total? (4)

D: Formal methods – addition and subtraction

About this topic: When calculations are too difficult to do mentally, you can use formal written methods.

Think back to the place value of numbers. When you add or subtract larger integers, it is helpful to write the numbers in their place value columns. Start with the units column, then the tens column and so on. Revise these skills by looking through the worked examples. Then repeat each one without looking at the method and answer.

Addition

Addition of integers with carrying

	Th	H	T	U
		3	0	8
+	1	9	7	5
	2	2	8	3
			1	1

308 + 1975

Line up the units digits and write down any, carried digits below the answer line.

8 + 5 = 13   Write down 3 in the units column and carry the 1 ten to the tens column.

	TTh	Th	H	T	U
	2	5	6	7	8
		6	4	3	9
			2	8	8
+				7	9
	3	2	4	8	4
	1	1	2	3	

25678 + 6439 + 288 + 79

> Set out your calculations clearly and neatly to avoid making silly mistakes.

Notice the small 'carried' digits.

Addition of decimals with carrying

You use the same method to add decimal numbers. Remember to line up the units digits in the same column.

	Th	H	T	U	•	t	h	th
		1	0	5	•	3	7	
				0	•	1	7	4
+		2	0	9	•	0	6	
		3	1	4	•	6	0	4
				1		2		

105.37 + 0.174 + 209.06

Write each number so that the decimal points align.

Subtraction

You can also use a frame when you subtract numbers. Start with the column on the right and work to the left.

Subtraction of integers without 'borrowing'

	H	T	U
	2	6	9
−	1	2	5
	1	4	4

269 – 125

First, subtract 5 units from 9 units.

Then subtract 2 tens from 6 tens.

Finally, subtract 1 hundred from 2 hundreds.

Subtraction of integers with 'borrowing'

If the top number in a column is smaller than the bottom number, 'borrow 1' from the column on the left and add it to the top number.

TTh	Th	H	T	U
4	⁷8̶	¹1	⁴5̶	¹6
−	1	6	5	3	9
	3	1	6	1	7

48 156 − 16 539

Subtracting 9 units from 6 units gives a negative result, so change 1 ten into 10 units.

Then subtract 9 units from 16 units and 3 tens from 4 tens.

There are several ways of showing 'borrowing'. Practise using your preferred method.

Subtraction of decimal numbers with 'borrowing'

The same rules apply when subtracting decimal numbers. Line up decimal points carefully.

U	•	t	h
²3̶	•	¹⁰1̶	¹0
− 0	•	6	7
2	•	4	3

3.10 − 0.67

Subtracting 7 hundredths from 0 hundredths gives a negative result.

Change 1 of the tenths to 10 hundredths: 10 − 7

Change 1 unit into 10 tenths: 10 − 6

U	•	t	h
0	•	6	7
+ 2	•	4	3
3	•	1	0

Check your answer by using the inverse. You can do this in your head.

If there are zeros in the top line, keep moving to the left until you find the next non zero number.

Train

1 Calculate these using formal methods. Show all your working, including carried digits. (6)
   (a) Add 143 to 37
   (b) Subtract 34 from 111
   (c) 5.7 + 6.9
   (d) Subtract 2.48 from 13.5
   (e) 1345 + 5687
   (f) 6798 − 2342

Try

2 Show all your working, including carried numbers. (4)
   (a) Which number is 5.6 less than 12.3?
   (b) What must be added to 11.4 to get 23.2?
   (c) 89 568 + 5643 + 211 + 56
   (d) 18 001 − 9398

Test

3 What is the difference between four thousand and seven and one thousand, three hundred and sixty-five? (1)

4 In an election, 25 678 voted for the blue party, 13 551 voted for the purple party and 789 for the yellow party.
   (a) How many people voted altogether? (1)
   (b) How many more votes did the blue party get than the yellow party? (1)

E: Short multiplication and multiplication by multiples of 10

When multiplications are too difficult to do mentally, you can use formal written methods.

As with formal addition and subtraction methods, set out formal multiplications in columns.

Short multiplication of integers

Th	H	T	U
	4	1	8
×			7
2	9	2	6
		1	5

418×7

$7 \times 8 = 56$

Write the 6 in the units column and carry 5 into the tens column.
Note the small carry digit.

Short multiplication of decimals

When multiplying decimals, leave out the decimal point when you set out the calculation.
There are two methods for putting the decimal point into the answer to the first calculation.

Th	H	T	U
	1	1	5
×			6
	6	9	0
		3	

11.5×6

Method 1
Notice there is one digit (5) to the right of a decimal point in the question.

This means there must be one digit to the right of the decimal point in the answer. The answer is 69.0

Method 2
You can work out the position of the decimal point by approximating the answer to the calculation.

$11 \times 6 = 66$ and $12 \times 6 = 72$, so the answer must be between 66 and 72

The answer is 69.0

Multiplication by 10, 100 and 1000

To multiply a number by 10, move the digits one place to the left.

TTh	Th	H	T	U	
		4	6	7	× 10
	4	6	7	0	

$467 \times 10 = 4670$

To multiply a number by 100, move the digits two places to the left.

To multiply a number by 1000, move the digits three places to the left.

TTh	Th	H	T	U	
			5	6	× 1000
5	6	0	0	0	

$56 \times 1000 = 56\,000$

The same rules apply to decimal numbers. Any empty columns on the right are filled with zeros, up to and including the units digit.

Th	H	T	U	t	h	th	
			4 •	0	9		× 1000
	4	0	9 •				

$4.09 \times 100 = 409$

Multiplication by multiples of 10, 100 and 1000

When you multiply by **a multiple** of 10, 100 or 1000, do the calculation in stages.

$$102 \times 300 = 102 \times 100 \times 3$$
$$= 10\,200 \times 3$$
$$= 30\,600$$

Remember to estimate before you calculate.

	M	HTh	TTh	Th	H	T	U
				6	7	3	2
×				4	0̸	0̸	
	2	6	9	2	8	0	0
		2	1				

6732×400

Estimate $7000 \times 400 = 2\,800\,000$

First cross out the zeros and drop them down to the answer line.

Then multiply column by column.

Finally, compare your answer with your estimate.

Train

1 Show your working. (6)
   (a) 19×8 (c) 671×4 (e) 2561×100
   (b) 56×6 (d) 1795×9 (f) $2.567 \times 100\,000$

Try

2 Multiply: (a) 245×7 (b) 1943×60 (c) 4222×8000 (3)
3 What is the total cost of 5 DVDs at £8.75 each? (1)
4 Which of these multiplications has the largest result? (5)
   A: 24×5 B: 25×4 C: 54×2 D: 42×5 E: 52×4
5 Consider these three multiplications: 67×8 86×7 78×6
   (a) Without working out the answers, which calculation will give: (3)
       (i)   the smallest result   (ii) the largest result.
   (b) Work out the actual answers and compare them with your answers to part (a). (3)

Test

6 8630×7000 (1)
7 An airport bus carries 156 people. It makes 40 journeys in a week and is always full. How many people does it carry during a week? (1)
8 A chef is ordering flour to make cakes. He needs 200 g per cake and plans to make 560 cakes. How much flour should he order? (Give your answer in grams.) (2)

F: Multiplication by a two-digit number

About this topic: Factors and partitioning can be used when you need to multiply by a two-digit number. Long multiplication is a more formal method.

All the methods are designed to make the multiplications easier.
You will need to know your times tables, so keep practising them!

For all methods:

- Estimate first, so you have a rough idea of what to expect.
  For example, 123 × 24 will give a result of around 2500 (100 × 25)
- Write down what you are doing at each stage of the calculation.
- Check that the result is sensible at each stage. Use what you know to check your answers.
  For example, a number ending in 4 multiplied by a number ending in 3 always gives a number ending in 2 because 4 × 3 = 12

Multiplication by factors

Any composite (non-prime) number can be written as a **product of its factors** (see Chapter 1).

When multiplying by a two-digit number such as 24 or 32, it is often simpler and quicker to multiply by factors. Think of all possible products of factors first and then choose the one that makes the calculation easiest.

Multiply 189 by 32

200 × 30 = 6000	**Estimate** by rounding the numbers being multiplied.
32 = 4 × 8 or 32 = 4 × 4 × 2 or ...	Choose a suitable product of factors.
189 × 32 = 189 × 4 × 8	Multiply by your chosen factors in turn.
= 756 × 8	Use a frame for both multiplications if you find it easier.
= 6048	
Check the units digit of the result.	9 × 2 = 18, so the result will have a units digit 8 ✔

Multiplication by partitioning

Calculate 468 × 24

500 × 20 = 10 000   **Estimate** by rounding the numbers being multiplied.

This tells you that you will need TTh and Th columns as well as HTU.

Partition the number you are going to multiply by into tens and units and rewrite the calculation:
468 × 24 = (468 × 20) + (468 × 4)

Multiply the numbers in the brackets.

	TTh	Th	H	T	U
			4	6	8
×				2	Ø
		9	3	6	0
		1	1		

	TTh	Th	H	T	U
			4	6	8
×					4
		1	8	7	2
			2	3	

Then **add** the two products.

	TTh	Th	H	T	U
		9	3	6	0
+		1	8	7	2
	1	1	2	3	2
		1	1		

Long multiplication

Long multiplication is similar to partitioning, but the three steps are combined into one frame.

Calculate 1762 × 68   **Estimate:**   2000 × 70 = 140 000

	HTh	TTh	Th	H	T	U	
			1	7	6	2	
×					6	8	
		1	4⁶	0⁴	9¹	6	×8
+	1	0⁴	5³	7¹	2	0	×60
	1	1	9	8	1	6	
				1			

×8 First multiply by 8 (Note how the carry numbers are written.)

×60 Then multiply by 60 (Remember to fill the unit space with a zero.)
Add the products. (Write the carry numbers from the addition below the answer line.)

Remember to check your answer by comparing it to the estimate.

> **Key words:** products of factors, partitioning, long multiplication

Train

1 Match each calculation with the correct answer.   (4)

218 × 16	6176
302 × 28	3488
193 × 32	5768
412 × 14	8456

Try

2 Use factors to complete these multiplications.   (4)
    (a) 45 × 12   (b) 122 × 15

3 Use partitioning to complete these multiplications.   (G)
    (a) 512 × 28   (b) 1640 × 37

4 Use long multiplication to complete these multiplications.   (6)
    (a) 246 × 21   (b) 895 × 49

Test

For each of these questions, use a method of your choice.

5 37 × 24   (2)

6 512 × 28   (2)

7 Flights to Norway cost £197 for adults and £146 for children. What is the total cost for 14 adults and 36 children?   (3)

G: Short division, division by factors and multiples of 10

About this topic: Revise methods for dividing numbers, including how to deal with remainders.

Short division

Always begin by dividing the digit with the highest place value and work across to the lowest place value. Work through the examples and then try them for yourself.

Calculate $5245 \div 5$ $5 \div 5 = 1$

	Th	H	T	U
		1	0	4	9
5	5	2	²4	⁴5

5 doesn't go into 2, so write 0 on the answer line and write 2 as a small carry digit

$24 \div 5 = 4$, remainder 4, so write 4 on the answer line and write 4 as a small carry digit

$45 \div 5 = 9$

When a number does not divide exactly into another number, there is a **remainder**.

Calculate $200 \div 7$

	H	T	U	
	0	2	8	
7	2	²0	⁶0	r 4

$20 \div 7 = 2$, remainder 6

$60 \div 7 = 8$, remainder 4

The answer is 28, remainder 4

You could carry on dividing until you have an answer to 2 decimal places.

	H	T	U	.	t	h	
	0	2	8	.	5	7	
7	2	²0	⁶0	.	⁴0	⁵0	r 0.01

The answer is 28.57 (2 d.p.)

In **problem-solving** questions, you need to decide how to deal with a remainder sensibly.

A chocolate factory makes 1426 chocolates. A box contains 6 chocolates. How many boxes can the factory pack?

$1426 \div 6 = 237$ r 4   Only full boxes count, so round the answer down.

The factory can pack 237 boxes.

1426 people are divided into groups of 6 as far as possible. How many groups are there?

$1426 \div 6 = 237$ r 4   Everyone must have a group, so round up.

There will be 238 groups.

Dividing by factors

When dividing by a two-digit number such as 24 or 36, it is often simpler to divide by factors.

Calculate $192 \div 24$. Choose a suitable product of factors: $24 = 2 \times 12$ or $24 = 2 \times 2 \times 6$ or ...

$192 \div 24 = 192 \div 2 \div 2 \div 6$ Divide by your chosen factors in turn.

$= 96 \div 2 \div 6$

$= 48 \div 6$

$= 8$

Dividing by multiples of 10

Calculate $540 \div 60$

Choose a suitable product of factors: $60 = 6 \times 10$

$540 \div 60 = 540 \div 10 \div 6$ 　　　　　　　or 　　　　　　　$54\cancel{0} \div 6\cancel{0} = 54 \div 6$

$\qquad\qquad = 54 \div 6$ 　　　　　　　　　　　　　　　　　　$= 9$

$\qquad\qquad = 9$

Calculate $260\,000 \div 13\,000$

Choose a suitable product of factors: $13\,000 = 13 \times 1000$

$260\,000 \div 13\,000 = 260\,000 \div 1000 \div 13$ 　or 　　　$260\,000 \div 13\,000 = 260 \div 13$

$\qquad\qquad\qquad = 260 \div 13$ 　　　　　　　　　　　　　　$= 20$

$\qquad\qquad\qquad = 20$

> **Key words:** remainder, factors, decimal place

Train ●

1　Calculate the following. Some answers may have remainders. 　　　　(4)

(a) $654 \div 3$ 　　　(b) $2468 \div 4$ 　　　(c) $1465 \div 6$ 　　　(d) $8845 \div 9$

Try ●

2　Calculate these by dividing by factors. 　　　　(6)

(a) $396 \div 36$ 　　　　　(c) $7644 \div 28$ 　　　　　(e) $45\,000 \div 500$

(b) $408 \div 6$ 　　　　　(d) $2176 \div 32$ 　　　　　(f) $240\,000 \div 6000$

Test ●

3　$72 \div 6 = 12$ Use this fact to write down answers to these questions. 　(4)

(a) $720 \div 6$ 　　　(b) $72 \div 60$ 　　　(c) $7.2 \div 6$ 　　　(d) $7.2 \div 12$

4　The village pond holds 5346 litres of water. A bucket can hold 3 litres. How many times must one bucket be filled, to empty the water from the pond? 　(2)

5　There are 12 inches in a foot. How many feet are there in 1152 inches? 　(2)

H: Long division

About this topic: The formal method of long division is useful for more complex division calculations. You will also consider remainders as fractions.

Remember to divide from the digit with the highest place value and work across to the lowest place value.

● Write down what you are doing at each stage of the calculation.
● Estimate each step and then check by multiplying. Write this next to your working.
● Take care to write all the answer digits in the correct places.

In short division, you divided the digits and carried over any remainders into the next column.

	Th	H	T	U
	0	3	2	1
8	2	²5	¹6	8

Calculate 2568 ÷ 8?

Estimate: 2600 ÷ 8 = 300

Short division calculations can be set out in a long division format.

	Th	H	T	U
	0	3	2	1
8	2	5	6	8
		2	4	
			1	6
			1	6
			0	8
			0	8
			-	-

Divide: 25 ÷ 8 = 3 r 1   Check: 8 × 3 = 24

Subtract: 25 − 24 = 1 and pull 6 down

Divide: 16 ÷ 8 = 2   Check: 8 × 2 = 16

Subtract: 16 − 16 = 0 and pull 8 down

Divide: 8 ÷ 8 = 1   Check: 1 × 8 = 8

Subtract: 8 − 8 = 0

This method is particularly useful when dividing by 2-digit numbers.

Calculate 5688 ÷ 36

Estimate: 5688 ÷ 36 ≈ 6000 ÷ 40 = 150

		Th	H	T	U
		0	1	5	8
3	6	5	6	8	8
		3	6		
		2	0	8	
		1	8	0	
			2	8	8
			2	8	8
			-	-	-

56 ÷ 36   Estimate 60 ÷ 40 = 1.5   Check 36 × 1 = 36

Subtract and pull down.

208 ÷ 36   Estimate 200 ÷ 40 = 5   Check 36 × 5 = 180

Subtract and pull down.

Divide 288 ÷ 36   Estimate 300 ÷ 40 = 7.5   Check 36 × 8 = 288

Compare with the estimate.

More about remainders

In 2G you dealt with remainders by writing them as, for example $200 \div 7 = 28 \text{ r } 4$
or dividing into decimal places, for example $200 \div 7 = 28.57$

28.57 is not exact the exact answer. You can write remainders *exactly* as fractions in their **lowest terms**.

	H	T	U	
	0	2	8	
7	2	^{2}0	^{6}0	r 4

Calculate $200 \div 7$

Because we are dividing by 7, the remainder 4 represents four-sevenths $\left(\frac{4}{7}\right)$

$200 \div 7 = 28\frac{4}{7}$ This is an exact answer.

	H	T	U	
	0	5	2	
6	3	^{3}1	^{1}5	3

Calculate $315 \div 6$

The remainder 3 represents three-sixths $\left(\frac{3}{6}\right)$ This is written as $\left(\frac{1}{2}\right)$ in its lowest terms.

$315 \div 6 = 52\frac{1}{2}$

> **Key words:** estimate, long division, lowest terms

Train

1   Set out each question as a long division, then calculate. (6)
   (a) $580 \div 4$ (c) $2401 \div 7$ (e) $888 \div 24$
   (b) $1655 \div 5$ (d) $432 \div 12$ (f) $972 \div 54$

Try

2   Use long division to solve these. Some have remainders. (6)
   (a) $410 \div 6$ (c) $545 \div 5$ (e) $5901 \div 42$
   (b) $2180 \div 32$ (d) $4758 \div 39$ (f) $6024 \div 36$

Test

Use your preferred method to answer these questions. Decide on the best way to give any answers with remainders.

3   Divide four hundred and eighty by twenty-four. (1)

4   I spent £55 on cupcakes. One cupcake costs 44p. How many cupcakes did I buy? (1)

5   A school bus can take 72 passengers. The whole school is going on a trip: there are 30 teachers and 1200 pupils. How many buses are needed? (2)

6   £962 is to be shared equally between seven charities. How much will each charity receive? (1)

Test 2

Remember to set out your calculations clearly and neatly to prevent losing valuable marks by making avoidable mistakes.

Show all your working.

1 Add 253 to 47   (1)

2 Subtract 44 from 123   (1)

3 4.6 + 7.9   (1)

4 Subtract 3.49 from 14.5   (1)

5 1045 − 647   (1)

6 1998 + 1996 + 966   (1)

7 Which number is 5.4 less than 14.4?   (1)

8 What must be added to 12.3 to get 24.2?   (1)

9 (a) Multiply 29 by 7   (b) 14 000 ÷ 700   (c) Multiply 456 by 14   (3)

10 (a) Divide 2045 by 5   (b) Divide 2002 by 7   (c) 2325 ÷ 9   (3)

11 How many of these calculations have the answer 8?   (2)

$3 \times 4 - 4$	$5 + 6 \div 2$	$4 \times 5 - 3$	$5 \times 6 - 2 \times 11$
$42 - 4 \div 2$	$42 - 23$	$4 \times (5 - 3)$	$(8 - 2)^2 - 7 \times 22$

   A: 3   **B:** 4   **C:** 5   **D:** 6   **E:** 7

12 (a) $24 + (12 \times 2) \times 5$ (1)

   (b) $42 + 5 \times (6 - 3) \div 5$ (1)

13 It takes 7 hours and 55 minutes to fly from London to New York and 6 hours 15 minutes to fly from New York to Los Angeles. How many minutes flying time is that altogether?   (1)

14 Share £40 between seven people as equally as possible.

   (a) How much will each person receive?   (1)

   (b) How much will be left over?   (1)

15 A cat's home orders 5000 cans of cat food a month. Each cat eats 20 tins of food a month. How many cats can they feed?   (2)

16 Chairs are set out in the school hall for speech day. 1456 people are expected to attend. There are 26 chairs in a row. How many rows are needed?   (2)

Record your score and time here and at the start of the book.

Score ☐ / 25   Time ☐ : ☐

③ Fractions, proportions and percentages

Introduction

You probably use fractions in everyday life without even noticing. 'It's half past seven'. 'We can share the pizza between us: we'll have a quarter each.' 'I walk a mile and a half to school.'

Before you start working through this chapter, remind yourself of what the words in bold mean.

Fractions describe the number of parts out of a whole. The **denominator** (bottom number) is the number of equal parts into which the whole or group is divided. The **numerator** (top number) tells us how many of those parts we are considering.

The value of a **proper fraction** is between zero and one. In a proper fraction, the numerator is smaller than the denominator, for example, $\frac{1}{2}, \frac{3}{4}, \frac{15}{20}$

The value of an **improper fraction** is greater than one. In an improper fraction, the numerator is larger than the denominator, for example, $\frac{3}{2}, \frac{6}{4}, \frac{11}{6}$

A **mixed fraction** has a whole-number part and a proper-fraction part, for example, $1\frac{1}{4}, 1\frac{3}{5}, 4\frac{1}{3}$

Proportion is a way of describing part of something in relation to the whole. For example, the proportion of the population that is left-handed is about one in every eight.

A **percentage** is the number of parts out of 100, for example, 50% is 50 parts out of 100

The work in the previous chapters related to the first rung of the maths Learning Ladder. This chapter covers fractions, ratio and proportion, the second rung of the maths Learning Ladder.

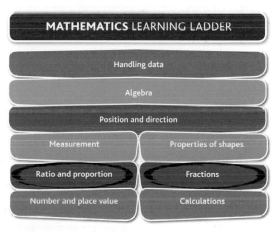

MATHEMATICS LEARNING LADDER

Handling data

Algebra

Position and direction

Measurement

Properties of shapes

Ratio and proportion

Fractions

Number and place value

Calculations

Advice for parents

There are many practical activities to reinforce the concept of fractions, for example, cutting up pizzas or cakes into equal parts and dividing them among family and friends. These are particularly useful for kinaesthetic learners. A kinaesthetic learner is someone who learns by doing something active rather than listening (auditory) or watching (visual).

Playing games with fractions is another way to support learning. Make a set of number cards with the numbers from 1 to 9. Shuffle the cards and deal four cards to both players. Take turns to make up challenges, such as:

* Who can make the largest proper fraction?
* Who can make the smallest improper fraction (see page 42)?
* Who can make a fraction that can be simplified?

When you have run out of ideas for challenges, shuffle the cards and deal them out again.

A: Equivalent fractions, improper fractions and mixed numbers

About this topic: Revise the basics of fractions including how to calculate equivalent fractions, improper fractions and mixed numbers, and how to write fractions in their simplest form (lowest terms). These terms crop up in many areas of mathematics. Think back to Chapter 2 when you wrote remainders as fractions in their lowest terms.

Equivalent fractions

Equivalent fractions have the same value but different **denominators** (bottom number).

The fractions $\frac{3}{4}$ and $\frac{6}{8}$ are equivalent fractions.

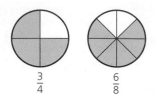

$\frac{3}{4}$ $\frac{6}{8}$

You can find an equivalent fraction by **multiplying (or dividing)** the top and bottom of a fraction by the **same number**.

$\frac{3 \times 3}{4 \times 3} = \frac{9}{12}$ $\frac{3}{4}$ and $\frac{9}{12}$ are also equivalent fractions.

$\frac{6}{12}$

It is straightforward to **compare** two fractions with the same denominator. If fractions have different denominators, you will need to find equivalent fractions with the same denominator before you can compare them. Use the **lowest common denominator** if you can. This is the **lowest common multiple (LCM)** of the two denominators.

Which is larger, $\frac{3}{4}$ or $\frac{2}{3}$?

Change into equivalent fractions with the same denominator. LCM of 4 and 3 is 12

$\frac{3}{4} = \frac{3 \times 3}{4 \times 3} = \frac{9}{12}$

$\frac{2}{3} = \frac{2 \times 4}{3 \times 4} = \frac{8}{12}$

$\frac{3}{4} \left(\frac{9}{12} \right)$

$\frac{2}{3} \left(\frac{8}{12} \right)$

$\frac{3}{4}$ is the larger fraction (by $\frac{1}{12}$)

You can use this method to **order** fractions (see 3B).

Writing fractions in their lowest terms

To write fractions in their lowest terms, reverse the process in the examples above. **Divide** the numerator and the denominator by the same number to find a simpler equivalent fraction. Divide by the **highest common factor (HCF)** of the numerator and the denominator if you can. If not, you will need to divide more than once.

$\frac{12}{20} = \frac{12 \div 2}{20 \div 2} = \frac{6}{10}$

$\frac{6}{10} = \frac{6 \div 2}{10 \div 2} = \frac{3}{5}$

$\frac{3}{5}$ cannot be simplified further.

It is in its simplest form (lowest terms).

Improper fractions and mixed numbers

In an **improper fraction**, the number of parts is more than one whole so the numerator is larger than the denominator. Improper fractions are often called **top-heavy fractions**.

$1\frac{1}{12}$ is an example of a **mixed number**. A mixed number has a whole number part and a proper fraction part. Mixed numbers can be changed into improper fractions and vice-versa.

$\frac{13}{12}$ is equivalent to 1 whole ($\frac{12}{12}$) and $\frac{1}{12}$

We can write this as $1\frac{1}{12}$

$1\frac{1}{2} = \frac{2}{2} + \frac{1}{2} = \frac{3}{2}$

$\frac{9}{4} = \frac{4}{4} + \frac{4}{4} + \frac{1}{4} = 2\frac{1}{4}$

> **Key words:** equivalent fraction, denominator, numerator, simplest form, lowest terms, improper fraction, top-heavy fraction, mixed number, lowest common multiple, lowest common denominator

Train

1 In this strip of squares, five out of seven have been shaded to show the fraction $\frac{5}{7}$

Sketch a similar diagram to show the fraction $\frac{5}{9}$ (1)

2 Are the two fractions in each pair equivalent? (4)
   (a) $\frac{1}{2}$ $\frac{3}{6}$ yes   (b) $\frac{9}{12}$ $\frac{3}{4}$ yes   (c) $\frac{12}{16}$ $\frac{3}{4}$ yes   (d) $\frac{3}{4}$ $\frac{12}{24}$ no

3 Write these fractions in their lowest terms. (4)
   (a) $\frac{2}{6}$ $\frac{1}{3}$ (b) $\frac{10}{20}$ $\frac{1}{2}$ (c) $\frac{16}{18}$ $\frac{4}{6}$ (d) $\frac{65}{180}$ $\frac{1}{6}$

Try

4 Fill in the missing numerator or denominator in each equivalent fraction pair. (6)
   (a) $\frac{1}{4} = \frac{4}{16}$ (b) $\frac{7}{15} = \frac{14}{30}$ (c) $\frac{8}{9} = \frac{40}{45}$ (d) $\frac{2}{3} = \frac{10}{15}$ (e) $\frac{5}{6} = \frac{20}{24}$ (f) $\frac{3}{5} = \frac{15}{25}$

Test

5 Write these proper fractions in their lowest terms. (3)
   (a) $\frac{9}{18}$ $\frac{1}{2}$ (b) $\frac{15}{25}$ $\frac{3}{5}$ (c) $\frac{9}{12}$ $\frac{3}{4}$

6 Write these improper fractions as mixed numbers. (3)
   (a) $\frac{9}{4}$ $2\frac{1}{4}$ (b) $\frac{12}{5}$ $2\frac{2}{5}$ (c) $\frac{17}{3}$ $5\frac{2}{3}$

7 There are 12 cakes in a box. 9 of them are chocolate cakes. What fraction of the cakes is this? Give your answer in its lowest terms. (2)

8 Ten of the 32 children in Year 6 are boys. What fraction of the class are girls? (2)

B: Ordering, adding and subtracting fractions; fraction sequences

About this topic: Now that you have revised the basics, practise ordering, adding and subtracting fractions.

Ordering fractions

We can put fractions in order of size by comparing them. To compare fractions, first change them to **equivalent fractions** with a **common denominator**.

Write $\frac{5}{6}, \frac{2}{3}$ and $\frac{3}{4}$ in order of size, starting with the smallest.

Find the LCM of the denominators by writing the numbers as a product of their prime factors. See 1D for a reminder about this.

$6 = 2 \times 3$ $3 = 3 \times 1$ $4 = 2 \times 2$ $3 \times 2 \times 2 = 12$ The LCM is 12

Calculate equivalent fractions: $\frac{5}{6} = \frac{5 \times 2}{6 \times 2} = \frac{10}{12}$ $\frac{2}{3} = \frac{2 \times 4}{3 \times 4} = \frac{8}{12}$ $\frac{3}{4} = \frac{3 \times 3}{4 \times 3} = \frac{9}{12}$

Write the fractions in order of size, smallest first: $\frac{2}{3}, \frac{3}{4}, \frac{5}{6}$

Adding fractions

It is straightforward to add fractions with the same denominator: just add the numerators.

$\frac{1}{3} + \frac{2}{3} = \frac{3}{3} = 1$

To add fractions with different denominators, first change them to **equivalent fractions** with a common denominator.

$\frac{3}{4} + \frac{2}{3} = \frac{9}{12} + \frac{8}{12} = \frac{17}{12} = 1\frac{5}{12}$

Subtracting fractions

It is straightforward to subtract proper fractions with the same denominator: just subtract the numerators.

$\frac{3}{5} - \frac{1}{5} = \frac{2}{5}$

To subtract fractions with different denominators, first change them to equivalent fractions with a common denominator.

What is the difference between $\frac{3}{4}$ and $\frac{1}{3}$?

$\frac{3}{4}$ and $\frac{1}{3}$ can be changed into equivalent fractions with a common denominator 12:

$\frac{3}{4} = \frac{3 \times 3}{4 \times 3} = \frac{9}{12}$ 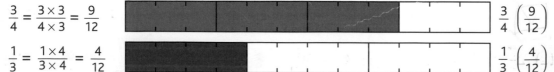 $\frac{3}{4} \left(\frac{9}{12}\right)$

$\frac{1}{3} = \frac{1 \times 4}{3 \times 4} = \frac{4}{12}$ $\frac{1}{3} \left(\frac{4}{12}\right)$

The diagrams clearly show that $\frac{3}{4}$ is larger and the difference between them is $\frac{5}{12}$

You can use the same method for mixed numbers.

Calculate $1\frac{1}{2} - \frac{1}{3}$

$1\frac{1}{2} = 1\frac{3}{6}$ 6 is the LCM of 2 and 3 and $\frac{1}{3} = \frac{2}{6}$

$1\frac{3}{6} - \frac{2}{6} = 1\frac{1}{6}$

If the fraction part in the number you are subtracting is bigger than the fraction part of the mixed number you are subtracting from, first change the mixed number into an improper fraction.

Calculate $4\frac{1}{6} - 1\frac{3}{4}$

$4\frac{1}{6} = 4\frac{2}{12}$ and $1\frac{3}{4} = 1\frac{9}{12}$ Find LCM, write as equivalent fractions with common denominator.

$4\frac{2}{12} = \frac{50}{12}$ and $1\frac{9}{12} = \frac{21}{12}$ $\frac{9}{12} > \frac{2}{12}$, so convert to improper fractions.

$\frac{50}{12} - \frac{21}{12} = \frac{29}{12} = 2\frac{5}{12}$ Subtract improper fractions.

Fraction sequences

Use the same methods for finding missing terms in fraction sequences as you do for integer and decimal sequences.

Find the next two terms in the sequence $\frac{1}{2}, 1\frac{3}{4}, 3, ..., ...$

$\frac{2}{4}, \frac{7}{4}, \frac{12}{4}, ..., ...$ Convert the fractions to equivalent fractions with the LCM 4

$\frac{12}{4} - \frac{7}{4} = \frac{5}{4} = 1\frac{1}{4}$ Calculate the difference between terms.

The next two terms are: $\frac{12}{4} + \frac{5}{4} = \frac{17}{4} = 4\frac{1}{4}$ and $\frac{17}{4} + \frac{5}{4} = \frac{22}{4} = 5\frac{1}{2}$

> **Key words:**
> lowest common multiple, equivalent fractions, common denominator

If you can't spot the pattern easily, you can convert to equivalent fractions, as shown in this example. There is no need to convert if you can spot the pattern.

Train

1   Write each set of fractions in order of size, largest first.   (3)

    Hint: find the lowest common denominator and calculate equivalent fractions.

    (a) $\frac{1}{2}$ $\frac{4}{7}$ (b) $\frac{8}{12}$ $\frac{7}{9}$ (c) $1\frac{1}{4}, \frac{16}{6}$ and $1\frac{2}{3}$

2   Calculate the following. Give your answers in their lowest terms.   (4)

    (a) $\frac{1}{5} + \frac{2}{5}$ (b) $1\frac{1}{6} + \frac{7}{12}$ (c) $\frac{7}{10} - \frac{1}{5}$ (d) $4\frac{4}{7} - 1\frac{1}{3}$

Try

3   Calculate: (a) $2\frac{3}{4} + 3\frac{4}{15} + 4\frac{1}{6}$ (b) $4\frac{1}{5} - 1\frac{9}{10}$ (2)

Test

4   Calculate the following. Give your answers in their lowest terms.   (2)

    (a) $3\frac{5}{6} - 1\frac{4}{5}$ (b) $\frac{6}{15} + 1\frac{2}{3}$

5   Find the missing terms in these patterns. Write fractions in their lowest terms.   (4)

    (a) $\frac{1}{3}, 1, 1\frac{2}{3}, ..., ...$ (b) $..., 1\frac{5}{16}, 1\frac{3}{8}, ..., 1\frac{1}{2}$

6   In the school triathlon, pupils run $4\frac{1}{2}$ miles, swim $\frac{3}{4}$ miles and cycle $6\frac{2}{3}$ miles. What is the total length of the race?   (2)

C: Ratio

About this topic: Fractions (such as $\frac{2}{3}$) compare the number of parts we have (in this case 2) to the number of parts in the whole (in this case 3). In this section, you will revise ratios which also compare amounts. You will meet this often in daily life and they can help you solve problems.

There are 15 beads on this string: 6 pink and 9 blue. ●●●●●●●●●●●●●●●

You can use fractions to talk about the beads, for example, $\frac{6}{15}$ of the beads are pink (or $\frac{2}{5}$ in its lowest terms).

You can also use ratios to express information about the beads.

The order of the numbers in ratios is very important.

	Colour 1	Colour 2	Total beads	Ratio
pink beads : total beads	6 pink		15	6:15 3:5 (simplest form)
pink beads : blue beads	6 pink	9 blue		6:9 2:3 (simplest form)
blue beads : pink beads	9 blue	6 pink		9:6 3:2 (simplest form)
total beads : pink beads : blue beads				15:6:9 5:2:3 (simplest form)

To write a ratio in its simplest form, divide the two parts of the ratio by the highest common factor.

The ratio of pink beads to blue beads can be written as 2:3

There are 12 tables and 24 benches in a dining room. What is the ratio of tables to benches?

Ratio of tables to benches = 12:24

$\qquad\qquad$ = 1:2

Problem solving with ratios

Work through the example of how to solve a problem involving ratios.

A florist has 108 tulips and 144 daffodils.

(a) What is the ratio of tulips to daffodils?
  Ratio of tulips to daffodils = 108:144 $\qquad$ Simplify the ratio by dividing both sides by 36

$\qquad\qquad$ = 3:4

(b) The florist wants to make up identical bunches of flowers that each contain 12 tulips. If she uses all the flowers, how many daffodils will there be in each bunch?
  From part (a), the ratio of tulips to daffodils is 3:4

$\quad$ 3:4 = 12:☐ $\qquad$ (12 ÷ 3 = 4)

$\qquad$ = 12:16 $\qquad$ (4 × 4 = 16)

There will be 16 daffodils in each bunch.

It is a useful to work out the **total number of parts** in a ratio and the **value of each part**. You can then check that all the parts add up to the whole.

Elderflower cordial is mixed with water in the ratio 1:4 to make elderflower juice. How much cordial is needed to make 1 litre of elderflower juice?

Ratio of elderflower cordial to water to juice = 1:4:5

$$= \square : \square : 1000 \quad 1 \text{ litre} = 1000 \, ml$$

$$= 200 : 800 : 1000 \quad 1000 \div 5 \text{ (total parts) is } 200 \text{ (one part)}$$

$$200 \times 4 = 800$$

200 m*l* of elderflower cordial is needed to make 1 litre of elderflower juice.

> **Key words:** ratio, highest common factor, lowest terms

Train

1 Write each ratio in its lowest terms. **(a)** 9:12 **(b)** 4:18 **(c)** 15:40 (3)

2 Write down the ratio of blue squares to green squares in each pattern. (2)

(a)

(b)

Try

3 Belle has 3 rabbits and 7 hamsters.

(a) What is the ratio of the number of hamsters to the number of rabbits? (1)

(b) What is the ratio of the number of hamsters to the total number of pets? (1)

4 A chef mixes 24 kg of strawberries with 6 kg of sugar to make jam.

(a) What is the ratio of the mass of strawberries to the mass of sugar? (1)

(b) If he has 32 kg of sugar, how many kilograms of strawberries will he need? (1)

(c) If he has 100 kg of strawberries, how much sugar will he need? (1)

Test

5 Mini ice creams are sold in large boxes and small boxes. Each box contains chocolate ice creams and mint ice creams in the ratio 5:4

(a) There are 10 chocolate ice creams in the small box. How many mint ice creams are in the box? (1)

(b) What is the total number of ice creams in a small box? (1)

(c) A large box contains 54 ice creams. How many of each type of ice creams are there? (1)

6 Holly, Ellie and Marianne share a box of biscuits in the ratio 3:2:4

(a) Marianne gets 12 biscuits. How many biscuits does each of the others get? (1)

(b) How many biscuits were in the box in total? (1)

(c) Holly eats a third of her biscuits and Ellie and Marianne each eat half of theirs. What is the ratio of the numbers of biscuits that Holly, Ellie and Marianne have left? Write your ratio in its simplest form. (2)

D: Proportion, equivalent measures, ratio and scale

About this topic: In this section, you will revise another way of describing part of something in relation to the whole.

Proportion

A **proportion** can be expressed as a **fraction** or a **percentage**.

- A school has 300 pupils of which 200 are boys; the proportion of pupils that are boys is two-thirds.
- Air is made up of nitrogen (78.09%), oxygen (20.95%), argon (0.93%), carbon dioxide (0.03%) together with water vapour and rare gases; the proportion of nitrogen in the air is 78.09%
- If a car is travelling at 50 miles per hour then:
  - it will travel 50 miles in 1 hour
  - it will travel 200 miles in 4 hours.
- A recipe lists ingredients for 12 buns. If you want to make more than 12 buns, you will need to increase the quantities of the ingredients in the correct proportions.

Work through the examples which show one method for solving problems with proportion.

Makes 20 muffins
250 g self-raising flour
125 g olive oil
250 ml semi-skimmed milk
2 eggs
200 g sugar
100 g chocolate chips

(a) How much flour is needed to make 100 muffins?
250 g for 20 muffins

$250 \div 20 = 12.5$ g of flour for 1 muffin

$12.5 \times 100 = 1250$ g of flour for 100 muffins

(b) I have 5 eggs. How many muffins can I make?
2 eggs make 20 muffins

1 egg makes 10 muffins $(20 \div 2)$

5 eggs make 50 muffins (10×5)

Equivalent measures

You can use the same method to compare equivalent measures.

5 miles is equivalent to 8 km. What distance in kilometres is equivalent to 20 miles?

5 miles = 8 km

1 mile = $\frac{8}{5}$ km

20 miles = $\frac{8}{5} \times 20 = 32$ km

A distance of 32 km is equivalent to 20 miles.

Ratio and scale

Plans and maps are drawn to a **scale**. The scale is usually given as a ratio.

A model of an office building has been built to a scale of 1:50

What actual length does 15 cm on the model represent?

scale:actual = 1:50   Write the scale.

= 1 cm:50 cm   Write the units.

= 15 cm:☐cm   Write the dimensions you know.

= 15 cm:750 cm   Calculate the missing dimension (50 × 15)

15 cm on the model represents 750 cm on the office building.

> **Key words:** proportion, fraction, percentage, equivalent measures

Train

1   A distance of 5 miles is equivalent to 8 km. What distance, in kilometres, is equivalent to: (3)
    (a) 10 miles   **(b)** 65 miles   **(c)** 160 miles?

2   A mass of 22 lb is equivalent to 10 kg. What mass, in pounds, is equivalent to: (2)
    (a) 5 kg   **(b)** 15 kg?

Try

3   These are the ingredients to make 12 pancakes.   (3)

3 eggs
115 g plain flour
1 teaspoon baking powder
140 ml milk

(a) How many eggs will I need to make 36 pancakes?

(b) I have 350 ml of milk. How many pancakes can I make?

(c) How much flour will I need to make 48 pancakes?

Test

4   In a school of 500 pupils, there are 300 boys. What proportion of the school is girls? Give your answer as a fraction in its lowest terms.   (1)

5   In a chemistry lesson, Zara mixes 18 mg of chemical X with 30 mg of chemical Y.
    (a) What is the ratio of the mass of chemical Y to the mass of chemical X?   (1)
    (b) If Zara has 108 mg of chemical X in total, how many mixtures can she make?   (1)
    (c) How much of chemical Y will she need to do this?   (1)

6   A model of a carriage is made to a scale 1:50
    (a) If the model is 24 cm long, what is the length of the actual carriage?   (1)
    (b) If the actual carriage is 425 cm wide, what is the width of the model?   (1)

E: Calculating with money and other decimal calculations

About this topic: You have done some revision of decimal numbers in Chapters 1 and 2. In this section, you will revise calculating with money and revisit other decimal calculations.

Calculating with money

When you are **adding** and **subtracting** money, it can be easier to convert everything to pence first. You can convert your answer back to pounds.

£2.32 + £5.07 = 232p + 507p = 739p = £7.39

£10 − 75p = 1000p − 75p = 925p = £9.25

If you need to add more amounts of money (or decimal numbers) than you can manage in your head, set up a grid (see 2C). Take care to align the numbers in the correct columns and remember to put the decimal point back in your answer.

Calculate £1.65 × 5

Use short multiplication:

£1.65 × 5 = £8.25

U	•	t	h
1	•	6	5
×			5
8	•	2	5
		3	2

There are two digits to the right of the decimal point of an amount in pounds so, when you multiply by a whole number, there will be two digits to the right of the decimal point in the answer.

When you **divide** an amount of money by an integer, convert to pennies first and then convert the answer to pounds.

£5 ÷ 4 = 500p ÷ 4

Use short division:

Convert to pounds: 125p = £1.25

H	T	U
	1	2	5
4	5	¹0	²0

More decimal calculations

Here is a reminder of the **place values** in decimal numbers.

In the number 1.257, 2 is two-tenths ($\frac{2}{10}$), 5 is five-hundredths ($\frac{5}{100}$) and 7 is seven-thousandths ($\frac{7}{1000}$)

H	T	U	•	t	h	th
100	10	1	•	$\frac{1}{10}$	$\frac{1}{100}$	$\frac{1}{1000}$
		1	•	2	5	7

- When **multiplying by 10, 100 or 1000**, the digits move to the left.

  2.34 × 10 = 23.4   The digits have moved 1 place to the left.

  0.006 × 1000 = 6   The digits have moved 3 places to the left.

- When **dividing by 10, 100 or 1000**, the digits move to the right.

  2.34 ÷ 10 = 0.234   The digits have moved 1 place to the right.

  6 ÷ 1000 = 0.006   The digits have moved 3 places to the right.

- You may be asked to **round** decimal numbers to a certain number of decimal places. As with whole numbers:
  - **Round down** if the value of the digit to the right of the 'cut-off' is 4 or below.
  - **Round up** if the value of the digit to the right of the 'cut-off' is 5 or more.

Round 0.1196 to 2 d.p.

The digit to the right of the 2nd decimal place is 9

(9 > 4), so round up.

Rounded to 2 d.p. 0.1196 is 0.12

● **Multiplying and dividing decimals by a one-digit number** (see 2E and 2G)

H	T	U	.	t	h	th
	6	8	.	1	2	
×					4	
2	³7	2	.	4	8	

	U	.	t	h	th
	0	.	6	2	8
8	5	.	⁵0	²2	⁶4

■ 68.12 × 4 = 272.48   ■ 5.024 ÷ 8 = 0.628

● **Multiplying and dividing decimals by a two-digit number** (see 2F and 2H)

	H	T	U	.	t	h	th
			1	.	9	9	8
×						1	7
		1	3⁶ .	9⁶	8⁵	6	×7
+		1	9 .	9	8	0	×10
		3	3 .	9	6	6	
			1	1	1		

■ 1.998 × 17 = 33.966

		T	U	.	t	h	th	tth
		0	0	.	5	8	2	
2	6	1	5	.	1	3	2	
		1	3	0				
			2	1	3			
			2	0	8			
					5	2		
					5	2		
					-	-		

■ 15.132 ÷ 26 = 0.582

> When the two-digit number is a multiple of 10, remember to write the 0 in the answer line.

Key words: decimal places, round up, round down, formal addition and subtraction

Train

1   Calculate: **(a)** £1.24 + £5.78   **(b)** £6.25 – 70p   **(c)** £14.02 × 3   **(d)** £31.50 ÷ 6   (3)

2   Copy and complete this shopping list.   (3)

2 pizzas at £4.25 each	....
1 tomato and basil sauce	£1.60
3 packets of tortellini at £1.90 each	....
Total	....

Try

3   **(a)** 15.15 × 100   **(b)** 2.75 ÷ 1000   **(c)** 6.45 × 20   **(d)** 12.126 ÷ 6   (4)

4   Round 2.7382 to: **(a)** 1 d.p.   **(b)** 2 d.p.   **(c)** 3 d.p.

Test

5   Here is the cost of Albie's shopping.   (2)

apples £1.25   carrots 70p   lettuce £1   potatoes £3.14   bananas 98p

(a) How much did he spend in total?   **(b)** How much change should he get from £20?

6   **(a)** 21.92 × 36   **(b)** 5.4 ÷ 72   (2)

7   Louis is saving up to buy a bicycle that costs £102. If he saves £12 a month, for how many months will he need to save before he can buy the bicycle?   (1)

8   When I add three numbers together and divide the result by 2, I get 8.678. If two of my numbers are 3.157 and 4.078, what is the third number?   (1)

F: Fractions and decimals

About this topic: You have already revised both fractions and decimals. In this section, you will revise the relationship between fractions and decimals.

Changing fractions to decimals

To change a fraction to a decimal:

- write it as an equivalent fraction with a denominator (bottom number) of 10, 100 or 1000
- write the numerator (top number) of the fraction with the correct place values.

$$\frac{1}{2} = \frac{1}{2} \times \frac{5}{5} = \frac{5}{10} \quad = 0.5$$

U	•	t
0	•	5

$$\frac{3}{5} = \frac{3}{5} \times \frac{2}{2} = \frac{6}{10} \quad = 0.6$$

U	•	t
0	•	6

$$\frac{3}{4} = \frac{3}{4} \times \frac{25}{25} = \frac{75}{100} = 0.75$$

U	•	t	th
0	•	7	5

You can also change a fraction to a decimal by dividing the numerator by the denominator.

	U	•	t			U	•	t			U	•	t	h
	0	•	5			0	•	6			0	•	7	5
2	1	•¹	0		5	3	•³	0		4	3	•³	0	²0

Use this method if you cannot change a fraction into an equivalent fraction with a denominator of 10 or any multiple of 10

$\frac{1}{3}$ cannot be changed to an equivalent fraction with a denominator 10, 100 or 1000, so divide 1 by 3

	0	•	3	3	3	3	3	...
3	1	•¹	0	¹0	¹0	¹0	¹0	...

This is an example of a recurring decimal.

We can write this as $0.\dot{3}$, where the dot over the $\dot{3}$ indicates that the 3 recurs.

Changing decimals to fractions

You can use place value to determine the value of a digit in a decimal number and therefore the equivalent fraction.

T	U	•	t	h	th
	1	•	5		
	6	•	0	8	
2	1	•	0	0	1
	0	•	0	0	5

1.5 is 1 and 5 tenths

6.08 is 6 and 8 hundredths

21.001 is 21 and 1 thousandth

0.005 is 5 thousandths

$1.5 = 1\frac{5}{10} = 1\frac{1}{2}$ in its simplest form

$6.08 = 6\frac{8}{100} = 6\frac{2}{25}$ in its simplest form

$21.001 = 21\frac{1}{1000}$ in its simplest form

$0.005 = \frac{5}{1000} = \frac{1}{200}$ in its simplest form

Write 12.045 as a mixed number.

T	U	•	t	h	th
1	2	•	0	4	5

$12\frac{45}{1000} = 12\frac{9}{200}$ in its lowest terms

Key words: numerator, denominator

Train

1 Write these fractions as decimals to 2 d.p. (4)

(a) $\frac{2}{3}$ 0.66   (b) $\frac{5}{6}$ 0.83   (c) $\frac{1}{9}$ 0.11   (d) $\frac{1}{40}$

2 Write these decimals as fractions in their lowest terms. (4)

(a) 0.25   (b) 1.6   (c) 11.001   (d) 12.456

Try

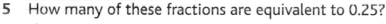

3 Write these fractions as decimals. (3)

(a) $\frac{28}{250}$ 8.92   (b) $\frac{18}{50}$ (c) $\frac{1}{7}$ (to 2 d.p) 0.14

4 Write these decimals as fractions in their lowest terms. (3)

(a) 4.27   (b) 12.004   (c) 0.010

Test

5 How many of these fractions are equivalent to 0.25? (2)

$\frac{1}{4}$ $\frac{4}{16}$ $\frac{20}{100}$ $\frac{75}{100}$ $\frac{25}{100}$ $\frac{1}{25}$ $\frac{12}{48}$

A: 7   B: 6   C: 5   D: 4   E: 3   F: 2   G: 1

6 Write these fractions as decimals.

(a) $\frac{40}{1600}$ (b) $\frac{3}{8}$ (2)

7 Write these decimals as fractions in their lowest terms.

(a) 0.35   (b) 1.372 (2)

8 My friend ate $\frac{5}{8}$ of a pizza.

(a) Write this as a decimal. (1)

(b) How much was left for me? Write your answer as a decimal. (1)

G: Percentages, decimals and fractions

About this topic: Now you have mastered changing fractions to decimals and decimals to fractions, you are going revise converting between fractions, decimals and percentages.

You have probably seen a 'percentage' complete bar on the screen when a computer is downloading files or completing updates.

Remember, a percentage is a number of parts out of 100

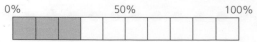

$$100\% = \frac{100}{100} = 1$$

Changing percentages to decimals and fractions

A percentage can easily be changed into a decimal or a fraction.

$$45\% = 0.45 = \frac{45}{100} = \frac{9}{20} \text{ in its lowest terms}$$

A percentage greater than 100% is equivalent to a fraction or decimal greater than 1

$$138\% = 1.38 = 1\frac{38}{100} = 1\frac{19}{50} \text{ in its lowest terms}$$

Sometimes a percentage has a fractional part. You can write this as a decimal fraction or a normal fraction.

$$75\frac{1}{2}\% = 75.5\% = 0.755 = \frac{755}{1000} = \frac{151}{200} \text{ in its lowest terms}$$

Changing decimals to percentages

To change a decimal to a percentage, multiply by 100

$$0.5 = 50\% \qquad 0.6 = 60\% \qquad 0.75 = 75\%$$

> Your answer is not complete without the % sign.

Changing fractions to percentages

A percentage is a fraction out of 100 so, to change a fraction to a percentage, first write it as an equivalent fraction with a denominator of 100

One way to do this is to multiply by $\frac{100}{100}$ and then cancel. In some cases you may be able to multiply by a smaller number to give an equivalent fraction with a denominator of 100

$$\frac{3}{5} \times \frac{100}{100} = \frac{300}{500} = \frac{60}{100} = 60\%$$

Some fractions cannot be changed into a fraction with a denominator of 10, 100, 1000 or any multiple of 10, for example $\frac{1}{3}$

To convert these fractions to percentages, divide the numerator by the denominator.

$\frac{1}{3}$ is written as 0.333... (0.$\dot{3}$) which is the same as $33\frac{1}{3}\%$

To write a mixed number as a percentage, first write it as an improper fraction.

$1\frac{9}{10} = \frac{19}{10}$ Write the mixed fraction as an improper fraction.

$= \frac{19}{10} \times \frac{100}{100} = \frac{1900}{1000} = \frac{190}{100}$ Multiply by 100 and cancel.

$= 190\%$

> **Key words:** percentage, fraction, decimal, lowest terms

Train ●

1   Write these percentages as: **(i)** decimals **(ii)** fractions in their lowest terms.   (6)

     (a) 5%   **(b)** 10%   **(c)** 55%

2   Write these as percentages.   **(a)** 0.1   **(b)** 0.42   **(c)** 0.98   (3)

3   Write these as percentages.   **(a)** $\frac{3}{4}$ **(b)** $\frac{2}{5}$ **(c)** $\frac{1}{40}$ (3)

Try ●

4   Write these percentages as: **(i)** decimals **(ii)** fractions in their lowest terms.   (6)

     (a) 97%   **(b)** 145%   **(c)** $60\frac{1}{2}\%$

5   Write these as percentages.   **(a)** 0.45   **(b)** 1.76   **(c)** 16.98   (3)

6   Write these as percentages.   **(a)** $1\frac{1}{10}$ **(b)** $2\frac{4}{25}$ **(c)** $1\frac{3}{20}$ (3)

Test ●

7   Copy and complete the table below to show equivalent fractions, decimals and percentages. (9)

Fraction (in simplest form)	$\frac{2}{5}$				$\frac{7}{10}$
Decimal		0.25		0.8	
Percentage	40%		35%		

8   Isla's marks in some tests are shown below.

     maths $\frac{87}{100}$ French 85%   biology 82%   English $\frac{44}{50}$ history $\frac{17}{20}$

     (a) What percentage did Isla score in: **(i)** maths **(ii)** English **(iii)** history?   (3)

     (b) In which subject did she get the highest score?   (1)

H: Recurring decimals; ordering fractions, decimals and percentages

About this topic: You should now be familiar with the relationships between fractions, decimals and percentages. In this section you will revise recurring decimals and how to order fractions, decimals and percentages.

Recurring decimals

Earlier in the chapter, you saw that when the fraction $\frac{1}{3}$ is changed to a decimal, it is a **recurring decimal**.

		0	3	3	3	3	3	...
3	1	·^{1}0	^{1}0	^{1}0	^{1}0	^{1}0		...

You can write this as: 0.33 (to 2 d.p.) or $0.\dot{3}$ where the dot indicates that the 3 recurs.

Look at what happens to these fractions when they are written as decimals.

$\frac{1}{7}$

		0	1	4	2	8	5	7	1	4	2	8	5	7	...
7	1	·^{1}0	^{3}0	^{2}0	^{6}0	^{4}0	^{5}0	^{1}0	^{3}0	^{2}0	^{6}0	^{4}0	^{5}0		...

We can write this as $\frac{1}{7} = 0.\dot{1}4285\dot{7}$ where the dots over the figures mark the start and end of the repeating unit.

$\frac{24}{99}$

			0	0	2	4	2	4	2	4	...
9	9	2	4	·0	^{42}0	^{24}0	^{42}0	^{24}0	^{42}0		...

We can write this as $\frac{24}{99} = 0.\dot{2}\dot{4}$ or 0.242 (3 d.p.).

Changing recurring decimals to percentages

To change a recurring decimal to a percentage, multiply by 100

0.3333...   0.3333... × 100 = 33.333...% = $33\frac{1}{3}$ %

0.5555...   0.5555... × 100 = 55.555...% = 55.56% (2 d.p.)

Ordering fractions, decimals and percentages

To order a set of numbers that contains fractions, decimals and percentages, first change them all to the same form.

Write these numbers in order, largest first.   75% $\frac{4}{5}$ 0.73

75% = 0.75 $\frac{4}{5}$ = 0.80   0.73

So, in order: $\frac{4}{5}$, 75%, 0.73

Key words: recurring decimal

Train

1 Write these fractions as recurring decimals. Use dots to show recurring digits. (3)

(a) $\frac{2}{3}$ (b) $\frac{1}{9}$ (c) $\frac{7}{33}$

2 (a) Write each of these recurring decimals as a percentage. (2)

(i) 0.4̇ (ii) 0.8̇3̇

(b) Write the decimals in part (a) as fractions in their lowest terms. (2)

Try

3 Write these fractions from the family of elevenths as decimals. Use dots over any recurring digits. (5)

(a) $\frac{1}{11}$ (b) $\frac{2}{11}$ (c) $\frac{3}{11}$ (d) $\frac{4}{11}$ (e) $\frac{5}{11}$

4 Write the family of sevenths as decimals. Give your answers to 3 d.p. (6)

5 Write the numbers in each set in order, smallest first. (2)

(a) 44% $\frac{12}{25}$ 0.49 $\frac{47}{100}$ (b) $\frac{55}{65}$ 0.87 85% $\frac{41}{50}$

Test

6 Copy and complete this table. (14)

Fractions	Decimal	Percentage
	0.5	
$\frac{1}{3}$		
	0.666...	
		25%
	0.75	
		20%
$\frac{1}{7}$		

7 Write the numbers in each set in order, largest first. (2)

(a) 0.208 $\frac{1}{4}$ 22% $\frac{1}{5}$ (b) 112% $1\frac{1}{9}$ 1.012 1.02 $1\frac{1}{12}$

I: Fractions of numbers and quantities

About this topic: Revise finding fractions of numbers or quantities.

Unit fraction of a number or quantity

A **unit fraction** has 1 as the numerator. To find a unit fraction of a number:

- divide the number by the denominator of the fraction
- or multiply the number by the fraction.

Calculate half of 72

$$72 \div 2 = 36 \quad \text{or} \quad 72 \times \frac{1}{2} = 36$$

If you cannot do the division mentally, use a written method.

Calculate $\frac{1}{8}$ of 20.48

$$\frac{1}{8} \text{ of } 20.48 = 20.48 \div 8 = 2.56$$

	T	U	•	t	h
		0	2	5	6
8	2	0	•^{4}4	^{4}8	

You can use the same method to find the unit fraction of a **quantity**.

Calculate one-quarter of 70 c*l*

$$70 \times \frac{1}{4} = 17.5$$

So one-quarter of 70 c*l* is 17.5 c*l*.

Calculate $\frac{1}{4}$ of £25

$$\frac{1}{4} \text{ of } £25 = £25 \div 4 = £6.25$$

Sometimes it can help to convert the units of the quantity to smaller units.

Calculate $\frac{1}{12}$ of 18 km

18 km = 18 000 m

		TTh	T	H	T	U
			0	1	5	0	0
	1	2	1	8	^{6}0	0	0

$$\frac{1}{12} \text{ of } 18\,\text{km} = 18\,000\,\text{m} \div 12 = 1500\,\text{m}$$

> Remember these conversions.
>
> 10 mm = 1 cm
> 1000 mm = 1 m
> 100 cm = 1 m
> 1000 m = 1 km
> 1000 mg = 1 g
> 1000 g = 1 kg
> 1000 m*l* = 1 litre

Non-unit fractions of a number or quantity

A **non-unit fraction** does not have 1 as a numerator.

To find a non-unit fraction of a number, divide by the denominator (to find one part) and then multiply by the numerator.

Calculate $\frac{3}{4}$ of 20

$$20 \times \frac{3}{4} = (20 \div 4) \times 3 = 5 \times 3 = 15$$

If you cannot do the calculations mentally, use a written method.

Calculate $\frac{3}{5}$ of 4.7 km

Work out $\frac{1}{5}$ of 4.7 km

4.7 km ÷ 5 = 0.94 km

		U	•	t	h
		0	•	9	4
5	4	•^{4}7	^{2}0		

Multiply the result by 3 to find $\frac{3}{5}$ of 4.7 km

0.94 km × 3 = 2.82 km

		U	•	t	h
		0	•	9	4
×					3
		2	•	8	2
		2		1	

> **Check that your answer is in the correct units.**

You also need to know how to **solve problems** involving fractions.

If a 1 m piece of string is cut into 5 equal pieces, how long will each piece be?

$\frac{1}{5}$ of 1 m = 100 cm ÷ 5 = 20 cm

Each piece of string will be 20 cm

Cancelling

The process of **cancelling** before you multiply can make calculations easier. This involves dividing a numerator and denominator in the calculation by a common factor.

Calculate $\frac{3}{4}$ of 60

$$\frac{\overset{15}{\cancel{60}} \times 3}{\underset{1}{\cancel{4}}} = 45$$

As before, it can be more straightforward to change the units before you calculate.

Calculate $\frac{3}{5}$ of 8 kg.

8 kg = 8000 g $\frac{3}{5}$ of 8000 g = $\frac{3}{\underset{1}{\cancel{5}}} \times \overset{1600}{\cancel{8000}}$ g = 4800 g or 4.8 kg

> **Key words:** unit fraction, non-unit fraction, common factor, cancelling

Train

1 Calculate these quantities. (4)

(a) $\frac{1}{2}$ of 1 m 500m   (b) $\frac{1}{3}$ of £100 33·33 (c) $\frac{1}{4}$ of 260 m 65m (d) $\frac{1}{8}$ of 260 km

(handwritten: 5⟌28, 33)

2 Calculate these quantities. (4)

(a) $\frac{2}{3}$ of 12 kg 8kg   (b) $\frac{3}{5}$ of £28 16·80   (c) $\frac{7}{10}$ of 18 litres 12·60   (d) $\frac{5}{8}$ of 2 km 1.25km

Try

3 The chef divides 2 kg of chocolate chips equally between 8 jars. How many grams of chocolate chips are in each jar? 250 kg (1)

4 Calculate these quantities. (4)

(a) $\frac{4}{9}$ of £2.70 1·20 (b) $\frac{3}{5}$ of 38 m 22·80 (c) $\frac{5}{12}$ of 1.8 km 0·75 (d) $\frac{2}{5}$ of £104.50 41·8

(handwritten: £56.25)

Test

5 Calculate these quantities.   (a) $\frac{1}{80}$ of £12 500   (b) $\frac{7}{9}$ of 54 litres   (2)

6 Calculate these.   (a) $\frac{24}{25}$ of 50   (b) $\frac{5}{36}$ of 180   (2)

7 A jug holds $\frac{3}{4}$ of a litre. How many millilitres is this? (1)

8 We travelled 451.5 km from London to Newcastle. At lunchtime, we had travelled $\frac{5}{6}$ of the total distance. How far was this? (1)

J: Finding the original amount

About this topic: Now you can find a fraction of an amount or quantity, revise finding the whole amount when you know how much a fraction of the amount or quantity is worth.

Finding an original amount with unit fractions

You already know how to find the unit quantity of a fraction. You can, for example, calculate that half of £72 is £36

Look at this information in a different way.

£36 is half the amount of money I have in my wallet. How much do I have in my wallet altogether?

You can approach this type of question by using n as the **unknown quantity**. In this instance, let n be the total amount of money I have.

Write the calculation as: $\frac{1}{2} \times n = £36$

So, $n \div 2 = £36$

Solve by multiplying. $n = £36 \times 2$ Multiplication is the opposite of division.

$n = £72$ The total I have in my wallet is £72

When a number is divided by 6, the answer is 4

What is the number?

Let n be the number: $n \div 6 = 4$

$$n = 4 \times 6$$

$$= 24$$ The number is 24

Finding an original amount with non-unit fractions

The method is similar to finding an original amount involving non-unit fractions, but you need to work through one additional step, as shown in the example below.

If $\frac{3}{4}$ of an amount is 24, what is the original amount?

Let n be the original amount, then: $\frac{3}{4} \times n = 24$

$3 \times n = 24 \times 4 = 96$ Multiply by 4

$n = 96 \div 3 = 32$ Divide by 3

Dividing before you multiply can simplify the calculation.

$$\frac{3}{4} \times n = 24$$

$\frac{1}{4} \times n = 24 \div 3 = 8$ Divide 24 by 3

$n = 8 \times 4 = 32$ Multiply by 4

You need to be able to use what you know to **solve problems**.

We have travelled 160 km. Mum says this is $\frac{2}{5}$ of the way to our campsite. What is the total distance we have to travel?

Let n be the total distance.

$$\frac{2}{5} \times n = 160\,km$$

$$\frac{1}{5} \times n = 160\,km \div 2 = 80\,km \quad \text{Divide by 2 to find } \frac{1}{5}$$

$$n = 80\,km \times 5 = 400\,km \quad \text{Multiply by 5 to find the total distance}$$

The total distance we have to travel is 400 km.

> Check that your answer is in the correct units.

> **Key words:** unit fraction, non-unit fraction, using n as the unknown quantity

Train

1 If $\frac{1}{2}$ of the total amount is 4, what is the total amount? (1)

2 If $\frac{1}{3}$ of the total amount is 12, what is the total amount? (1)

3 If $\frac{1}{7}$ of the whole is 10, what is the whole? (1)

4 If $\frac{2}{3}$ of a number is 18, what is the number? (1)

5 If $\frac{3}{7}$ of a number is 27, what is the number? (1)

6 If $\frac{5}{12}$ of a number is 60, what is the number? (1)

Try

7 When a number is divided by 3, the result is 11. What is the number? (1)

8 When a number is divided by 10, the result is 18. What is the number? (1)

9 If $\frac{5}{9}$ of a number is 45, what is the original number? (1)

10 When a number is divided by 3 and this result is multiplied by 8, the final answer is 24. What was the number? (1)

11 When a number is multiplied by 3 and this result is divided by 10, the final answer is 12. What was the number? (1)

Test

12 If $\frac{1}{11}$ of the total amount is 2, what is the total amount? (1)

13 If $\frac{7}{8}$ of the total amount is 14, what is the total amount? (1)

14 If $\frac{5}{6}$ of an amount of money is £35, what is the whole amount? (1)

15 When a number is divided by 2, the result is 105. What was the number? (1)

16 When our father shares out the money in his jar equally among four children, they each get £15. How much money was in the jar? (1)

17 If $\frac{7}{9}$ of a bag of sand has a mass 4.2 kg, what is the mass of a full bag of sand? (1)

18 A chef uses 45 kg of potatoes. This was $\frac{5}{6}$ of his total supply. What was the mass of his total supply? (1)

K: Percentage of a quantity and as a fraction

About this topic: You should be confident about the relationships between percentages, fractions and decimals. In this section you will revise calculating the **percentage of a quantity**.

A percentage is a number of parts out of 100

The whole is $\dfrac{100}{100} = 100\%$

The shaded blue square is $\dfrac{1}{100} = 1\%$

The green shaded area is $\dfrac{50}{100} = 50\%$

The red shaded area is $\dfrac{10}{100} = 10\%$

Percentages are useful when comparing, for example, results or amounts. The diagram clearly shows that 50% > 10% > 1%
Follow the method in the example below to **calculate a percentage** score.

Tom scored 9 out of 12 in a recent maths test. What is his score as a percentage?

$\dfrac{9}{12} = \dfrac{3}{4}$ Write 9 out of 12 as a fraction and simplify.

$\dfrac{3}{\underset{1}{4}} \times \overset{25}{100}\% = 75\%$ Multiply by 100%, cancelling first.

You can use logic to calculate simple percentages, such as 10% or 50%. You can also multiply by an equivalent fraction or decimal to work out a **percentage part of a quantity**.

Calculate 10% of 70 kg

$$70 \div 10 = 7$$

or $70 \times \dfrac{10}{100} = 7$

or $70 \times 0.1 = 7$

The answer is 7 kg

Calculate 50% of 3 litres

$$3 \div 2 = 1.5$$

or $3 \times \dfrac{1}{2} = 1.5$

or $3 \times \dfrac{5}{10} = 1.5$

or $3 \times 0.5 = 1.5$ The answer is 1.5 kg

Another method to find other percentage parts is to find 10% (or 50%) first.

Calculate 30% of £4.50

10% of £4.50 = 450p ÷ 10 = 45p First calculate 10%

$3 \times 10\% = 30\%$ Then multiply by 3 to calculate 30%

30% of £4.50 = 3 × 45p = 135p = £1.35

Calculate 55% of 3 tonnes

50% of 3 tonnes = 3 tonnes ÷ 2 = 1.5 tonnes

5% of 3 tonnes 1.5 tonnes ÷ 10 = 0.15 tonnes   Divide 50% by 10

55% of 3 tonnes   1.5 + 0.15 = 1.65 tonnes   Add 50% and 5%

The percentages we have considered so far are multiples of 5% and 10%. However, this is not always the case, for example, the average cost of heating a house increased by 12% this year. To calculate these percentages, first work out the value of 1% and then multiply.

Add 12% to an annual electricity bill of £205   1% of £205 = £205 ÷ 100 = £2.05

12% of £205 = £2.05 × 12 = £24.60

Adding this to the original bill, final bill = £205 + £24.60 = £229.60

Using fractions to calculate percentages

Sometimes the calculation is easier if you write the percentage as a fraction and then multiply.

Calculate $12\frac{1}{2}$% of 84 $\frac{12.5}{100} \times 84 = \frac{125}{1000} \times 84 = \frac{1}{8} \times 84 = 10.5$ Cancel to lowest terms before you multiply.

Calculate 9% of 300 $\frac{9}{100} \times 300 = 9 \times 3 = 27$

Any remainder can be written as a fraction or as a decimal.

> **Key words:** percentage of a quantity, percentage as a fraction, lowest terms

Train

1   Here are some recent test results. Work out the percentage score for each subject.   (4)

Subject	English	Mathematics	Science	History
Score	88 out of 100	56 out of 60	44 out of 50	65 out of 75

2   Calculate these percentages. Give your answers as integers or decimals, as appropriate.   (3)

   (a) 10% of 60   **(b)** 50% of 120   **(c)** 5% of 42

Try

3   There are 34 pupils in Year 1. 28 of them have dark hair. What percentage is this?   (1)

4   Flo was practising her goal shooting. Last week she scored 20 out of 25 shots. This week she scored 46 out of 50 shots. What is her percentage score rate for each week? Has she improved?   (2)

5   Calculate these percentages. Give your answers as integers or decimals, as appropriate.   (4)

   (a) 30% of 65   **(b)** 85% of 72   **(c)** 14% of 60   **(d)** $62\frac{1}{2}$% of 360 mg

Test

6   360 people were questioned before an election. 244 people said they would vote, 30 said they would not vote and the rest said they were undecided. Calculate the percentage of those asked that said they: (3)

   (a) would vote   **(b)** would not vote   **(c)** were undecided.

7   64% of a cinema audience was female.   (2)

   (a) What percentage of the audience was male?

   (b) 200 people watched the film. How many females were in the audience?

8   In a box of 72 sweets, 25% are mints, one-third are chocolates and the rest are toffees.   (3)

   (a) How many mints are there?   **(b)** How many toffees are there?

9   What is 17% of 225 mm?   (1)

10 What is 72% of 550 kg?   (1)

L: Percentage discounts and charges; profit and loss

About this topic: Many everyday events, including shopping, require a range of mathematical knowledge and skills. In this section you will revise how to calculate percentage discounts and charges (increases) and profit and loss.

Percentage discount

A **discount** on an item is often expressed as a percentage, for example 25% off.

After a discount: **original price − amount of discount = sale price**

The full price of a shirt is £30. It is reduced by 25%. What is the sale price?

Calculate the percentage discount: 25% of £30 $= \frac{25}{100} \times 30 = £7.50$

Then work out the sale price: £30 − £7.50 = £22.50

Percentage increase

After a price **increase**:   **original price + amount of increase = new price**

A restaurant adds a service charge of 12% onto the £65 cost of my food. What is the total bill?

Calculate the increase: 12% of £60 $= \frac{12}{100} \times 60 = £7.20$

Then work out the final price: £65 + £7.20 = £72.20

Profit or percentage increase

A **profit** is made when the **selling price** of an item is **greater** than the price it cost to buy it (the **cost price**). It is a **percentage increase** on the cost price. You might be asked to calculate the selling price or the **percentage profit**.

It costs a shop £25 to buy a pair of trousers cost from the wholesaler. They sell them at a profit of 50%. What is the selling price?

Calculate the profit: 50% of £25 $= \frac{50}{100}$ (or $\frac{1}{2}$) $\times £25 = £12.50$

Work out the selling price: £25 + £12.50 = £37.50   Selling price = original price + profit

A shopkeeper sells brooms for £12 each. They cost the shopkeeper £10 each. What is his percentage profit?

Calculate the profit amount: £12 − £10 = £2   Profit = selling price − cost price

Calculate the percentage profit: $\frac{2}{10} \times 100\% = 20\%$ Percentage profit $= \frac{profit}{cost\ price} \times 100$

Loss or percentage decrease

A **loss** is made when the **selling price** of an item is **less** than the **cost price**. It is a **percentage decrease** on the cost price.

The cost price of a set of speakers was £100. The shopkeeper made a loss of 20% when he sold them. What was the selling price?

Work out the amount of the loss: 20% of £100 $= \frac{20}{100} \times 100 = £20$

Calculate the selling price: £100 − £20 = £80   Selling price = cost price − loss

Calculate the percentage loss on a watch bought for £24 and sold for £16

Calculate the amount of the loss. £24 – £16 = £8

Work out the percentage loss: $\dfrac{8^{1}}{_{3}24} \times 100 = 33\dfrac{1}{3}\%$ Percentage loss = $\dfrac{\text{loss}}{\text{cost price}} \times 100$

The percentage loss was 33%

You can use the same methods to calculate other **percentage decreases**.

I trained for 20 hours last week, but only 15 hours this week. What is the percentage decrease in my training?

Work out the actual decrease: 20 hours – 15 hours = 5 hours

Work out the percentage decrease: $\dfrac{\text{decrease}}{\text{original hours}} \times 100 = \dfrac{5^{1}}{_{4}20} \times 100\% = 25\%$

Percentage decrease in training hours is 25%

> **Key words:** profit, loss, percentage charge/increase, percentage discount, selling price, cost price

Train

1   A pair of football boots cost £45 full price. They are reduced by 30%
    What is the reduced price? £30 (2)

2   The cost of a loaf of bread in my local shop last week was 80p. This week the price
    increased by 5%. How much does a loaf of bread cost now? 84p (?)

3   My travel agent charges 2% more for buying flights with a credit card. The cost of
    my flights without this charge was £560. What is the total cost with this charge? (2)

4   Twenty sausages are sold for £6. The cost price is £4.50. Calculate the percentage profit. (2)

Try

5   A computer was bought for £500 and sold for £320. What was the percentage loss? (2)

6   Value added tax (VAT) is currently 20%. I am given quotes for various items without VAT.
    Work out the actual price of each item when VAT is added. (3)
    (a) Paint, £20.00   **(b)** Wallpaper, £49.30   **(c)** Electrical work, £745

Test

7   The table shows the price of a
    remote-controlled car in four shops
    and their reductions in the sale. Copy
    and complete the table to show which
    shop is offering the best deal.

Shop	Pre-sale price	Discount	Sale price
A	£58	20% off	46.4
B	£62	Reduced by $\frac{1}{4}$	
C	£68	$\frac{1}{3}$ off all cars	
D	£70	30% off	£49

(4)

8   House prices have increased by 15% in the last year. What is the value of a house worth
    £85 000 one year ago after this increase? (1)

9   Yesterday the sun shone for 10 hours; today it shone for 6 hours. What is the percentage
    decrease in the hours of sunshine? (2)

M: Multiplying with fractions

About this topic: You will now revise multiplying a fraction by an integer, a mixed number by an integer or fraction, and a fraction by another fraction.

Multiplying a fraction by an integer

Look at these examples, which show you how to multiply a fraction by an integer.

$$2 \times \frac{2}{3} = \frac{2 \times 2}{3} = \frac{4}{3} = 1\frac{1}{3}$$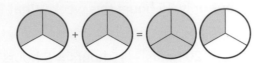

$$3 \times \frac{6}{10} = \frac{3 \times 6}{10} = \frac{18}{10} = 1\frac{8}{10} = 1\frac{4}{5}$$

Written as a mixed number in its lowest terms

Multiplying a mixed number by an integer

When you multiply a mixed number by an integer, multiply the integer part first and then the fraction part. Look at this example.

$$5 \times 2\frac{2}{3} = 5 \times 2 + 5 \times \frac{2}{3} = 10 + \frac{5 \times 2}{3} = 10 + \frac{10}{3} = 10 + 3\frac{1}{3} = 13\frac{1}{3}$$

Multiplying a fraction by another fraction

To multiply a fraction by another fraction, multiply the two numerators and multiply the two denominators.

> Remember that 'of' means multiply, so 'half of one-third' means $\frac{1}{2} \times \frac{1}{3}$

Follow the visual demonstration of this method.

The diagrams show a whole, a third and half of a third. You can see that half of a third is $\frac{1}{6}$

We can write the calculation:

$\frac{1}{2} \times \frac{1}{3} = \frac{1}{6}$ Multiplying the numerators gives the numerator of the result: $1 \times 1 = 1$
Multiplying the denominators gives the denominator result: $2 \times 3 = 6$

This method works for all proper fractions.

$$\frac{1}{4} \times \frac{1}{5} = \frac{1}{20}$$

$$\frac{2}{3} \times \frac{1}{4} = \frac{2}{12} = \frac{1}{6}$$ Expressed in its simplest form (lowest terms)

$$\frac{2}{3} \times \frac{3}{5} = \frac{6}{15} = \frac{2}{5}$$ Expressed in its simplest form (lowest terms)

Cancelling (dividing by a common factor) will make your calculations simpler.

$$\frac{\overset{1}{\cancel{4}}}{\underset{3}{\cancel{9}}} \times \frac{\overset{1}{\cancel{3}}}{\underset{2}{\cancel{8}}} = \frac{1}{6}$$

Multiplying a mixed number by a fraction

To multiply a mixed number by a fraction, first change the mixed number into an improper fraction.

$$\frac{1}{2} \times 3\frac{2}{3} = \frac{1}{2} \times \frac{11}{3} = \frac{11}{6} = 1\frac{5}{6} \qquad 4\frac{2}{3} \times \frac{9}{14} = \frac{\overset{1}{\cancel{14}}}{\underset{1}{\cancel{3}}} \times \frac{\overset{3}{\cancel{9}}}{\underset{1}{\cancel{14}}} = \frac{3}{1} = 3$$

> **Key words:** fraction, integer, mixed number, common factor, improper fraction

Train •

1   Work these out. Write your answers as mixed fractions in their lowest terms.   (6)

(a) $3 \times \frac{1}{2}$ (c) $2 \times \frac{1}{6}$ (e) $5 \times 1\frac{1}{4}$

(b) $5 \times \frac{4}{5}$ (d) $2 \times 2\frac{1}{5}$ (f) $4 \times 2\frac{3}{20}$

Try •

2   Work these out. Give your answers in their simplest form.   (6)

(a) $4 \times \frac{1}{6}$ (c) $\frac{3}{8} \times \frac{3}{5}$ (e) $4\frac{1}{2} \times \frac{1}{3}$

(b) $\frac{2}{3} \times \frac{1}{9}$ (d) $2\frac{2}{3} \times \frac{1}{5}$ (f) $1\frac{1}{5} \times \frac{2}{3}$

3   Work these out. Write your answers as mixed fractions in their lowest terms.   (6)

(a) $5 \times \frac{8}{12}$ (c) $2 \times \frac{7}{9}$ (e) $2 \times 1\frac{4}{5}$

(b) $4 \times \frac{5}{16}$ (d) $5 \times 3\frac{9}{10}$ (f) $4 \times 6\frac{1}{8}$

Test •

4   Work these out. Give your answers in their simplest form.   (9)

(a) $\frac{1}{2} \times \frac{1}{3}$ (d) $\frac{8}{15} \times \frac{5}{12}$ (g) $\frac{4}{9} \times 3\frac{5}{8}$

(b) $\frac{1}{3} \times \frac{3}{4}$ (e) $\frac{7}{8} \times \frac{4}{21}$ (h) $6\frac{2}{3} \times \frac{1}{4}$

(c) $\frac{5}{21} \times \frac{9}{10}$ (f) $1\frac{1}{2} \times \frac{3}{4}$ (i) $\frac{3}{7} \times 10\frac{1}{2}$

N: Dividing by fractions

About this topic: Now that you can multiply by fractions, you will revise different methods to divide by fractions.

Dividing an integer by a fraction

When you **divide an integer by a proper fraction**, the answer is larger than the original number.

I have one bar of chocolate which I divide among some friends. Each friend receives $\frac{1}{4}$ of a bar. How many friends do I share it among?

To answer this question, divide the integer 1 by the fraction $\frac{1}{4}$

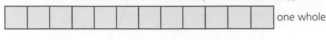

one whole

$\frac{1}{4}$

Four friends will each receive $\frac{1}{4}$ of the bar of chocolate.

I have two bars of chocolate. How many people could I give one-third of a bar of chocolate?

To answer this question, divide the integer 2 by the fraction $\frac{1}{3}$

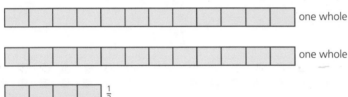

one whole

one whole

$\frac{1}{3}$

Six people could each be given one-third of a bar of chocolate.

To divide an integer by a fraction, rewrite as a multiplication:

- replace the division sign with a multiplication sign *and*
- write the **reciprocal** of the second fraction (turn it upside down).

So, dividing by $\frac{1}{3}$ is the same as multiplying by $\frac{3}{1}$

$$2 \div \frac{1}{3} \rightarrow 2 \times \frac{3}{1} = 6$$

Dividing a fraction by a fraction

Look at the examples of how to **divide a fraction by a fraction**.

You have half a bar of chocolate. How many people can you give a quarter of a chocolate bar?

The diagrams show that there are 2 quarters in a half.

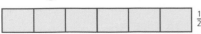

$\frac{1}{2}$

$\frac{1}{4}$

The calculation is $\frac{1}{2} \div \frac{1}{4} \rightarrow \frac{1}{2} \times \frac{4}{1} = \frac{4}{2} = 2$

You can give 2 people a quarter of a chocolate bar.

$$\frac{3}{4} \div \frac{1}{2} \rightarrow \frac{3}{4} \times \frac{2}{1} = \frac{6}{4} = 1\frac{1}{2} \qquad\qquad \frac{1}{2} \div \frac{3}{4} \rightarrow \frac{1}{2} \times \frac{4}{3} = \frac{4}{6} = \frac{2}{3}$$

Problem solving

It is best to think of problems in mathematics as puzzles or challenges. Read the problem carefully. It is important that you understand what the challenge is before you start trying to solve it.

Working through this example will help you with the final questions at the end of this section.

I have $\frac{1}{2}$ kg of raisins to <u>divide equally</u> among 6 people. What fraction of a kilogram will each friend receive?

Underline the important information if it helps you.

Then work out the calculation(s) that you will need to solve, in this case $\frac{1}{2} \div 6$

$$\frac{1}{2} \div 6 \rightarrow \frac{1}{2} \times \frac{1}{6} = \frac{1}{12} \qquad \text{So, each person will get } \frac{1}{12} \text{ kg of raisins.}$$

> **Key words:** dividing, integer, fraction, reciprocal

Train ●

1   Complete these divisions. If the answer is an improper fraction, write it as a mixed number in its lowest terms.   (6)

   (a) $2 \div \frac{1}{2}$ (b) $4 \div \frac{1}{6}$ (c) $\frac{1}{2} \div \frac{1}{3}$ (d) $\frac{2}{3} \div \frac{1}{6}$ (e) $1\frac{1}{2} \div \frac{3}{4}$ (f) $3\frac{1}{3} \div \frac{1}{12}$

Try ●

2   Complete these divisions. Give any improper fractions as mixed numbers in their lowest terms.   (6)

   (a) $2 \div \frac{2}{3}$ (b) $6 \div \frac{2}{5}$ (c) $\frac{1}{3} \div \frac{3}{4}$ (d) $\frac{6}{15} \div \frac{3}{10}$ (e) $3\frac{3}{10} \div \frac{6}{11}$ (f) $1\frac{5}{9} \div \frac{7}{12}$

Test ●

3   Work out these. Show your working clearly.   (6)

   (a) $4 \div \frac{10}{11}$ (b) $14 \div \frac{7}{9}$ (c) $\frac{9}{20} \div \frac{3}{4}$ (d) $\frac{3}{7} \div \frac{6}{21}$ (e) $1\frac{1}{2} \div \frac{2}{3}$ (f) $2\frac{3}{4} \div \frac{7}{8}$

4   The teacher has $\frac{3}{4}$ litre of orange squash to share between 15 children. What fraction of a litre does each child get?   (1)

5   The farmer had $\frac{7}{10}$ of a bucket of dry food. He gave $\frac{1}{4}$ of it to the sheep and the rest to the goats. What fraction of a bucket of dry food did

   (i) the sheep and   (ii) the goats each have?   (2)

Test time: 30:00

1 What fraction of this rectangle has been shaded? Give your answer
   in its simplest form. (1)

2 Change: (a) $\frac{1}{4}$ into twelfths (b) $\frac{2}{5}$ into twentieths. (2)

3 Write these fractions in order of increasing size. $\frac{2}{5}$ $\frac{3}{4}$ $\frac{1}{2}$ $\frac{2}{3}$ $\frac{1}{4}$ $\frac{5}{6}$ (1)

4 Write these mixed numbers as improper fractions. (a) $2\frac{3}{4}$ (b) $1\frac{19}{20}$ (2)

5 Write these improper fractions as mixed numbers. (a) $\frac{7}{2}$ (b) $\frac{27}{5}$ (2)

6 Calculate these. (a) $\frac{1}{2} + \frac{2}{3}$ (b) $1\frac{1}{2} - \frac{3}{5}$ (2)

7 To make hot chocolate, you add liquid chocolate to milk in the ratio 1:5

   (a) How much milk will I need to add to 500 ml of liquid chocolate? (1)

   (b) Altogether, I want to make 6 litres of hot chocolate. How much liquid
       chocolate will I need? (1)

8 I mix 1.2 kg of raisins, 0.35 kg of glacé cherries and 0.25 kg of nuts. I divide the mixture
   equally into 25 pots. How many grams of mixture is in each pot? (1)

9 Write these numbers in order, smallest first. 0.665 67% $\frac{13}{20}$ $\frac{2}{3}$ (1)

10 What is $\frac{3}{4}$ of £48? (1)

11 A vase holds $\frac{5}{6}$ litre of water. How many millilitres is this? (1)

12 If $\frac{3}{4}$ of an amount is 12, what is the original amount? (1)

13 A sand pit is being filled from a sack of sand. $\frac{3}{4}$ of the sack is 120 kg. What is
   the total mass of a full sack of sand? (1)

14 Sam's latest test scores for maths and science are: maths $\frac{37}{50}$ science $\frac{17}{25}$

   Calculate his percentage score for each subject. (2)

15 The cost price of a pencil case is £1.50. It is sold for a profit of 50%. What was its
   selling price? (1)

16 The cost price of a jacket was £25. It was sold for a loss of 12%. What was its
   selling price? (1)

17 Complete these calculations. Write each answer in its lowest terms. (2)

   (a) $3\frac{3}{5} \times \frac{1}{8}$ (b) $2\frac{5}{16} \div \frac{5}{8}$

18 A recipe for 36 muffins uses $1\frac{1}{2}$ cups of flour. How many cups of flour will I need to
   make 24 muffins? (1)

**Record your score and time here and
at the start of the book.**

Score [   ] / 25 Time [   ]:[   ]

Measures, shape and space

Introduction

Measures have been an important feature of everyday life since ancient times. Can you imagine a world without measures, shapes and space?

In mathematics we usually focus on plane shapes, solid shapes, symmetry, angles, patterns, maps and scale drawings. Pause for a moment and look around the room you are in. How many examples of these things can you see? As you look around, imagine you are making up questions to test your friends – what questions might you ask to test their knowledge?

When solving puzzles and problems about numbers or shapes, you may need to do the following:

- Use your knowledge and experience, make careful observations and think more than you write.
- Look for patterns and relationships.
- Ask yourself questions, such as 'What if...?' and ' How about...?'.
- Use a variety of strategies. Be prepared to try a different strategy if you are not succeeding.
- Present results in a clear and organised way, checking your results are sensible.

During this chapter, you will climb two rungs on the mathematics Learning Ladder. Once you have revised the skills in the first four chapters, you are over half way to the top of the ladder!

Problem solving

When a question is set in a real-life context, take your time to identify the mathematics behind the words. In this chapter you will learn formulae that help you calculate the perimeters and areas of shapes. They can help you find the missing pieces of information when you need to solve a problem. Working out how much fencing you need to surround a garden isn't as tricky as it might seem!

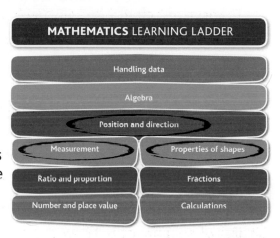

MATHEMATICS LEARNING LADDER

- Handling data
- Algebra
- Position and direction
- Measurement | Properties of shapes
- Ratio and proportion | Fractions
- Number and place value | Calculations

Advice for parents

As part of the Non-Verbal Reasoning tests, pupils will need to look for patterns and relationships that can involve shapes and numbers. Patterns in shapes can come from their properties, such as symmetry or faces, or by manipulating the shapes (reflection, rotation and translation).

Folding, turning and measuring the angles and sides of a range of cut-out shapes will give pupils a 'feel' for the shapes and measures they will meet in this chapter. This is particularly helpful for kinaesthetic learners. A kinaesthetic learner is someone who learns by doing something active rather than listening (auditory) or watching (visual).

Geometric puzzle activities, such as the one shown here, are useful to reinforce the concepts surrounding shapes and space. In a traditional tangram puzzle, a square is cut into seven geometric pieces which are then arranged in different ways to make countless different shapes.

Make up your own, similar puzzle. For example, you could start with a regular octagon and divide it into six geometric shapes. Cut out the pieces carefully and then work out how many different ways you can put them together to make interesting shapes. Think about the symmetry of each shape you make.

A: Reading number scales

About this topic: You will already be familiar with the scales on a variety of measuring instruments such as rulers and thermometers. In this section you will practise this important skill.

Measuring length and distance

Measuring instruments include: **rulers** for shorter straight lines, **tapes** to measure round your waist and **trundle wheels** to measure the distance around a running track.

You may need to choose which unit to use, for example, the distance between two towns is better measured in miles or kilometres than in centimetres!

How long is this beetle?

It is 0.9 cm long.

Units of measurement include:

Metric units	Imperial units
1 kilometre (km) = 1000 metres (m)	1 mile = 1760 yards (yd)
1 m = 100 centimetres (cm)	1 yd = 3 feet (ft)
1 cm = 10 millimetres (mm)	1 foot (ft) = 12 inches

To **convert** between metric and imperial units:

kilometres → ÷ 1.6 → miles   metres → ×(1.1)→ yards   centimetres → ×(0.4)→ inches

miles → × 1.6 → kilometres   yards → ÷(1.1)→ metres   inches → (÷ 0.4)→ centimetres

Try to remember these conversions:

8 km → 5 miles	1 m → 40 inches	30 cm → 1 foot	2.5 cm → 1 inch

Measuring mass

Measuring instruments include: **scales**, spring balance and microbalance (for very small masses).

Units of measurement include:

Metric units	Imperial units
1 tonne = 1000 kilograms (kg)	ounce (oz) pound (lb) stone (st)
1 kg = 1000 grams (g)	hundredweight (cwt) ton
1 g = 1000 milligrams (mg)	

To **convert** between metric and imperial units:

6.35 kg → 1 stone	1 kg → 2.2 lb	450 g → 1 lb

Measuring temperature

A **thermometer** is the most common instrument for measuring temperature. There are several different types: mercury thermometers, plastic thermometer strips (used on the forehead) and electronic thermocouples (used in the kitchen to check the temperature of foods).

Look at the scales on some thermometers. The units are often written in degrees Fahrenheit (°F) and degrees centigrade (°C). The centigrade scale is also known as the Celsius scale.

The table shows some useful temperatures in Fahrenheit and centigrade.

Body temperature	37 °C	98.6 °F
Freezer	⁻18 °C	0 °F
Fridge	3 to 5 °C	37 to 41 °F

Try to remember these conversions:

$0\,°C \rightarrow 32\,°F$ $100\,°C \rightarrow 212\,°F$

> **Key words:** distance, kilometre, metre, centimetre, millimetre, mile, tonne, kilogram, gram, thermometer, °C, °F

Train

1   List these lines in order of increasing length. Do not measure them.   (2)

A ———————————————————— 2

B ———————————————— 1

C ——————————————————————— 3

D ———————————————————————— 4

2   Choose appropriate units of length from the list below to measure each item.   (2)

mm   cm   m   km

(a) the thickness of an exercise book   (b) the distance between two railway stations

3   Choose appropriate units of mass from the list below to measure each item.   (2)

mg   g   kg   tonne

(a) the mass of a grain of sugar   (b) the mass of a large bulldozer

Try

4   At midday, the temperature was 11 °C; at midnight it had fallen to ⁻2 °C. By how many degrees had the temperature decreased?   13 °C   (1)

5   Lola's mass is 45 kilograms. Write Lola's mass in: (3)

(a) grams   (b) pounds   (c) stones and pounds.

Test

6   A snake is 3.05 metres long. Write this length in: (3)

(a) centimetres   30 5   (b) millimetres   (c) feet.

7   Last night, the temperature was ⁻4 °C. By lunchtime it had risen by 10 degrees. What was the temperature at lunchtime?   (1)

8   A chocolate cake uses these ingredients.

250 g butter   250 g sugar   350 g flour   550 g of chocolate chips   5 eggs (50 g each)

(a) Write the mass of each ingredient in kilograms.   (1)

(b) What is the total mass of all the ingredients in kilograms?   (1)

(c) If the cake loses 15% of its mass during baking, what is its mass after it is baked?   (1)

B: Working with time

About this topic: Time is another measure or scale. The most common units of time are seconds, minutes and hours.

Units of time

Try to learn the equivalences between the minutes and parts of hours below. They will be useful when you calculate with time.

1 hour	= 60 minutes		$\frac{1}{5}$ hour (0.2 h)	= 12 minutes
$\frac{1}{2}$ hour (0.5 h)	= 30 minutes		$\frac{1}{10}$ hour (0.1 h)	= 6 minutes
$\frac{1}{3}$ hour	= 20 minutes		$\frac{1}{60}$ hour	= 1 minute
$\frac{1}{4}$ hour (0.25 h)	= 15 minutes		1 minute	= 60 seconds

There are 24 hours in a day, 7 days in a week and 52 weeks in a year.

> You may need to know the number of days in each month to answer some questions. This rhyme will help you learn them.
>
> *30 days has September, April, June and November*
>
> *All the rest have 31*
>
> *Except February which has 28 days and 29 in each leap year.*

24-hour clock

Times can be written using either the 12-hour clock or the 24-hour clock. The table shows times written in both formats.

In the 12-hour clock, a.m. stands for *anti-meridiem* and means before noon. p.m. stands for *post-meridiem* and means after noon.

To change a 12-hour clock time to a 24-hour clock time, add 12 to the hours after 12.59 p.m. For a.m. times, add a 0 before a single-digit hour.

24-hour clock	12-hour clock
00:00 (midnight)	12.00 a.m.
07:30	7.30 a.m.
11:59	11.59 a.m.
12:00 (midday)	12.00 p.m.
12:59	12.59 p.m.
13:00	1.00 p.m.
16.05	4.05 p.m.
23:59	11.59 p.m.

It is important to understand units of time so that, for example, you can read bus and train **timetables** and understand different **times around the world**.

1.00 p.m. becomes 13:00   9 a.m. becomes 09:00

You do not need to write a.m. or p.m when using the 24-hour clock.

Time as a fraction

It is easier to calculate with time if you write both hours and minutes as **fractions of hours**. You can use what you have learned about mixed numbers and writing fractions in their lowest terms.

3 hours and 15 minutes = $3\frac{15}{60}$ hours = $3\frac{1}{4}$ hours

1 hour and 35 minutes = $1\frac{35}{60}$ = $1\frac{7}{12}$ hours

> **Key words:** 12-hour clock, 24-hour clock, fractions of hours, mixed number, lowest terms

Train

1 How many minutes are there in two hours? (1)

2 How many minutes are there in two days? (1)

3 These analogue clocks all show p.m. times. Write them as 24-hour clock times. (3)

(a)  20:40

(b)  23:59

(c) 12:16

4 Write these times as fractions of hours. (a) 20 minutes (b) 1 hour and 45 minutes (2)

Try

5 How many days are there between 26th September and 11th November (inclusive)? (1)

6 Write these 24-hour clock times as analogue times using 'past' or 'to'. (3)

   (a) 02:15   10   (b) 07:35   12   (c) 23:54   54

7 The time is 11:11 in London. Sydney (Australia) is 11 hours ahead. What time is it in Sydney?   2h and 35 (1)

8 Write these times using minutes and hours. (a) $\frac{1}{4}$ hour (b) $3\frac{2}{3}$ (c) $2\frac{5}{12}$ (3)

   15mn   3h and 40 man

Test

9 Spring term starts on 10th January and ends on 24th March. It is not a leap year. How many days, including both the first and last days and weekends, is it until the end of term? (1)   73days

10 I am travelling from Newcastle to Tunbridge Wells by train.

   (a) The train leaves Newcastle at 12:00 and arrives at London Kings Cross at 14:51
       How long does the journey take?   2 h and 51 min (1)

   (b) The underground train leaves Kings Cross at 15:06 and arrives at Charing Cross
       23 minutes later. At what time does it arrive at Charing Cross?   15:29 (1)

   (c) I finally arrive at Tunbridge Wells at 16:28   4 h and 28 min
       How long has my entire journey from Newcastle taken? (1)

11 Write 3 hours and 40 minutes as a fraction of hours. $3\frac{2}{3}$ (1)

12 Write $4\frac{5}{6}$ hours in hours and minutes.   4h and 50 min (1)

50

24
28
52
21
73

C: Speed, distance and time

About this topic: Building on the earlier work on time, revise the relationship between speed, distance and time and using the formula that connects them.

Calculating speed

The time taken for any journey will depend on the distance travelled and the speed of travel. The faster you travel, the shorter your journey time will be.

Units of speed include:

Metric units	**Imperial units**
metres per second (m/s)	miles per hour (miles/h *or* mph)
kilometres per hour (km/h)	

$1\,km/h \rightarrow 0.28\,m/s$ (the speed of a very fast mouse!)

Useful **conversions** to remember:

$30\,mph \rightarrow 48\,km/h \qquad 60\,mph \rightarrow 96\,km/h$

One method for calculating speed is to work out the distance travelled in one hour. If it takes 1 hour to travel 50 km, the average speed is 50 km/h.

Look at these examples.

A car takes 2 hours to travel 50 km. What is its average speed on the journey?

50 km in 2 hours, so divide by 2 to work out the distance travelled in 1 hour

$50\,km \div 2 = 25\,km \qquad$ The speed of the car on this journey is 25 km/h

A lorry travels 15 miles in 20 minutes.
What is its average speed on the journey?

15 miles in 20 minutes   1 hour = 60 minutes so multiply 15 and 20 by 3

45 miles in 60 minutes

The speed of the lorry on this journey is 45 mph

> *Give your answer in the correct units.*

Calculating distance

If you know the speed at which something is travelling and the time it takes, you can work out the distance travelled. Write the **time as a fraction** (proper or improper) and multiply it by the speed.

There is a useful formula for calculating speed, distance or time.

distance (d) = speed (s) × time (t)

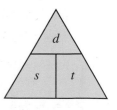

Louis cycled at a speed of 20 mph. How far did he travel in $1\frac{1}{2}$ hours?

Use the formula: $\qquad$ distance $= 20 \times 1\frac{1}{2} = \overset{10}{\cancel{20}} \times \dfrac{3}{\underset{1}{\cancel{2}}} = 30$ miles

> *If time is given in minutes, remember to write it in hours – as a fraction or decimal.*

Calculating time

The triangle for speed, distance and time can be used to calculate the time taken for a journey:

$$\text{time} = \frac{\text{distance}}{\text{speed}}$$

A bus travels 125 miles at a speed of 50 mph. How long does the journey take?

Use the formula: $\text{time} = \frac{\text{distance}}{\text{speed}} = \frac{125}{50}$ hours $= 2\frac{25}{50}$ hours $= 2\frac{1}{2}$ hours

The journey takes 2 hours 30 minutes.

> **Key words:** metric, imperial, time, distance, speed, formula

Train

1 Select the most likely mode of transport (car, bicycle or walking) for each set of results. (4)
   (a) distance 5 km, time 10 minutes   **(c)** distance 1.5 km, time 30 minutes
   (b) distance 230 m, time 6 minutes   **(d)** distance 1.6 km, 6 minutes

2 My Morris Minor travels 50 miles in one hour. What is its average speed? (1)

3 A racing car travels 60 miles in 30 minutes. What is its average speed? (1)

4 Isla rides her horse for 30 minutes at a speed of 6 km/h. What distance does she travel? (1)

Try

5 Amy lives twice as far from school as I do. We both leave for school at the same time.
   (a) If we walk at the same speed, who will arrive first? (1)
   (b) Amy scoots at twice the speed she walks. If she scoots to school, who arrives first? (1)

6 I run 2 km in 12 minutes. What is my average speed? (1)

7 It takes Ynes 20 minutes to cycle from home to school at a constant speed of 6 mph.
   How far is her school from her home? (1)

8 How long will it take to travel 12 miles at a speed of 36 mph? (1)

Test

9 Two cars travel from London to Cambridge. The red car travels at 70 mph and the blue car
   travels at 65 mph. Which car has the shorter journey time? (1)

10 A car travels a distance of 25 km in 15 minutes. Work out its average speed. (1)

11 Ben ran 200 metres in 5 minutes and Jasper ran 1.5 km in 20 minutes. Who ran faster? (2)

12 The Eurostar takes 2 hours to travel from London to Paris at an average speed
   of 170 km/h. What is the distance of the journey? (1)

13 A car leaves Cambridge at 10:00 and travels at a constant speed of 40 mph.
   At what time will it arrive in Oxford, a journey of 180 miles? (2)

D: Congruent shapes and similar shapes

About this topic: In this section, you will begin your revision of shapes, starting with congruent and similar shapes.

Congruent shapes

Two shapes (or objects) are congruent if they have the same shape and size. They might appear different at first if they have been turned (rotated) or flipped (reflected) but, if you put one on top of the other, they will be the same.

A piece of tracing paper is a useful tool to test whether or not shapes are congruent.

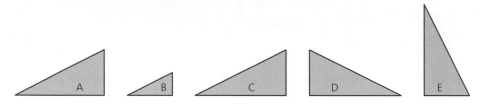

Triangles A and C are congruent – they are the same shape and size.

Triangle B is not congruent to any other triangle – it is smaller than A and C.

Although triangles D and E are in different positions, they are the same shape and size as A and C and so are also congruent to triangles A and C. Check with your tracing paper.

Similar shapes

Two shapes (or objects) are **similar** if they are exactly the same shape but *not* the same size. The lengths of the sides of the shape will be in the same **ratio**. This ratio is called the **scale factor**.

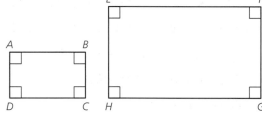

These two rectangles are the same shape but the sides of *EFGH* are twice the length of the corresponding sides of *ABCD*. Measure them to check.

We can say: $AB:EF = AC:EG$

The measurements show that the lengths of the sides of *ABCD* to *EFGH* are in the ratio 1:2 and, therefore, that the scale factor between the shapes is 2

When shapes are similar, you can use the scale factor to work out the lengths of missing sides.

In triangle *ABC* below, *AB* = 1.5 cm, *AC* = 2.5 cm and *BC* = 2 cm.

Work out the lengths of the unknown sides.

The triangles are similar, so the lengths of the sides will be in the same ratio.

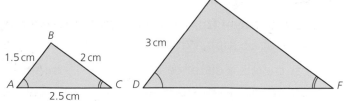

$AB:DE$ = 1.5:3 = 3:6 = 1:2 (simplest form)

The scale factor is 2. Use this to work out the lengths of the missing sides.

$EF = 2 \times BC = 2 \times 2 = 4$ $DF = 2 \times AC = 2 \times 2.5 = 5$

So, *EF* = 4 cm and *DF* = 5 cm

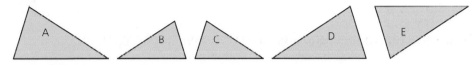

Write ratios using integers and not fractions
or decimals, for example 2 : 3 not 1 : 1.5

Key words: congruent, similar,
ratio, integers, scale factor

Train ●

You can use tracing paper for these questions.

1

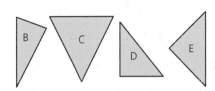

   (a) Which triangles are congruent to triangle A? (1)

   (b) Which triangles are congruent to triangle B? (1)

You will need a protractor.

2 Which triangles are similar?
   Measure the angles to check. (1)

Try ●

3

   (a) Which quadrilaterals are congruent to quadrilateral A? (1)

   (b) Which quadrilaterals are congruent to quadrilateral B? (1)

4 Triangles *ABC* and *DEF* are similar.

   (a) What is the ratio of *AB* to *DE*? (1)

   (b) What is the length of:

     (i) *DF* (1)

     (ii) *EF*? (1)

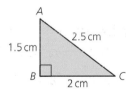 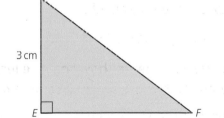

Test ●

5 Which of these shapes are congruent to each other? (2)

6 In a rectangle *ABCD*, *AB* = 5 cm and *BC* = 10 cm. Rectangle *WXYZ* is similar to rectangle *ABCD*, with *WX* = 15 cm.

   (a) Sketch the two rectangles. (1)

   (b) What is the length of *XY*? (1)

   (c) What is the ratio of the lengths of corresponding sides of rectangle *ABCD* to *WXYZ*? (1)

   (d) What is the scale factor? (1)

E: Plane shapes

About this topic: Revise the properties of regular and irregular two-dimensional shapes, including special triangles.

Plane shapes are flat or two-dimensional. They have **vertices** (corners), **sides** and **angles**. They may also have:

- **diagonals** – lines drawn across the shape from one vertex to another (not necessarily through the mid-point)
- one or more lines of **symmetry** (see section G)
- **rotational symmetry** about the mid-point (see section G).

Regular and irregular shapes

In a **regular** shape, all the angles are the same size and all the sides are the same length.

In an **irregular** shape, the angles and sides are not all the same size or length.

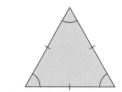
Regular (equilateral) triangle

Irregular triangle (scalene)

Regular pentagon

Irregular pentagon

Special triangles

Irregular triangles can have some equal angles and lengths.

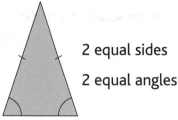

2 equal sides

2 equal angles

Isosceles triangle

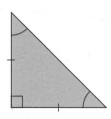

Right-angled triangle

2 equal sides

2 equal angles

Most right-angled triangles are scalene, but this one is an isosceles right-angled triangle.

Parallel lines

Parallel lines are lines that will never meet. They are marked with small arrows.

These quadrilaterals all contain at least one pair of parallel lines.

Parallel   Not parallel

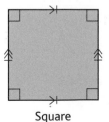
Square
All sides equal
All angles 90°
2 pairs parallel sides

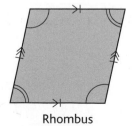
Rhombus
All sides equal
2 pairs equal angles
2 pairs parallel sides

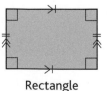
Rectangle
2 pairs of parallel equal sides
All angles 90°

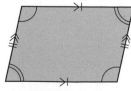
Parallelogram
2 pairs parallel equal sides
2 pairs equal angles

Key words: two-dimensional, vertex, angle, diagonal, regular, irregular, lines of symmetry, rotational symmetry

Train

1 Draw: (2)
   (a) a regular quadrilateral   (b) an irregular quadrilateral.

2 Copy the shapes below. (6)

(a)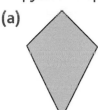

(b)

   (i) Mark any equal angles with arcs.
       (Use double arcs for a second pair of equal angles.)

   (ii) Mark any parallel lines with small arrows.
       (Use double arrows for a second pair of parallel lines.)

   (iii) Mark any pairs of equal sides with short lines.

Try

3 Copy the shapes below and mark any equal angles, parallel lines and equal sides as you did in question 2 (2)

(a) 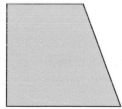   (b)

Test

4 (a) Draw a regular hexagon. What makes it regular? (4)
  (b) Draw an irregular hexagon. (2)

5 Copy the shapes below and mark any equal angles, parallel lines and equal lengths, as you did in question 2 (2)

(a) 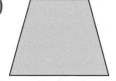   (b)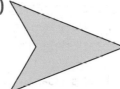

F: Symmetry

About this topic: In this section, you will revise reflection (line) symmetry and rotational symmetry.

Reflection symmetry

A shape or object has **reflection symmetry** if it can be folded in half so that the two sides match exactly.

The fold line is called the **line of symmetry**. On a drawing, the lines of symmetry are drawn as dotted lines.

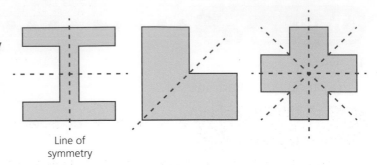

Line of symmetry

When you look in the mirror, you see your reflection. It is your **mirror image**. If you put a mirror on a line of symmetry of a shape, you will be able to see the missing half of the shape in the mirror. Try it on the shapes below.

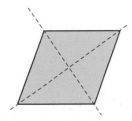

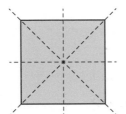

A square has 4 lines of symmetry. Notice that the corresponding lines of symmetry meet at right angles.

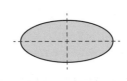

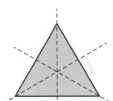

An equilateral triangle has 3 lines of symmetry.

In the drawing on the right, a design has been reflected in the dotted line to form a new shape.

The two halves of a shape, on either side of a line of symmetry, are the mirror images of each other. They are exactly the same shape and size which means the two halves are **congruent**.

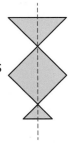

Rotational symmetry

A shape has **rotational symmetry** about its centre point if it can be rotated to fit onto itself in more than one way. The number of times it fits on itself in one complete turn is called the **order** of rotational symmetry.

You can test this by tracing a shape, marking the centre point, putting the point of your pencil on the centre point and then rotating the tracing. If the shape fits onto itself before you get back to the starting position, the shape has rotational symmetry. Try it with this square.

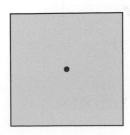

A square has a rotational symmetry of order 4

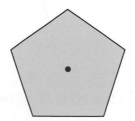

A regular pentagon has a rotational symmetry of order 5

4 Measures, shape and space

A shape can have rotational symmetry but no lines of symmetry.

This decorated square has rotational symmetry of order 4 but it has no lines of symmetry.

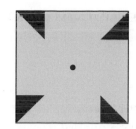

Key words: symmetry, line of symmetry, reflection symmetry, congruent, rotational symmetry

Train

1 Copy these shapes and draw on **all** lines of symmetry. (3)

(a)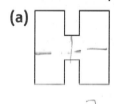

(b)

(c)

Try

2 Copy these shapes and mark on **all** lines of symmetry. How many lines of symmetry does each shape have? (3)

(a)

(b)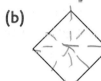

(c)

Test

3 How many of these shapes have rotational symmetry? (4)

square   kite   equilateral triangle   rhombus   rectangle
scalene triangle   isosceles triangle   parallelogram

A: 4   **B:** 5   **C:** 6   **D:** 7   **E:** 8

4 (a)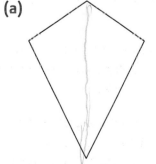

(b)

(c)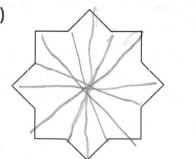

(9)

(i) Name each shape.

(ii) Copy each shape and draw all lines of symmetry. How many lines of symmetry does each shape have?

(iii) Describe the rotational symmetry of each shape.

G: Properties of quadrilaterals

About this topic: Quadrilaterals are flat shapes with 4 sides, 4 vertices and 4 angles. In this section you will revise the properties of different quadrilaterals.

Types of quadrilateral

You will come across quadrilaterals a lot in mathematics and in everyday life. Use these diagrams to revise the properties of some common quadrilaterals.

- **square**

all sides equal; all angles equal (90°); opposite sides parallel; diagonals equal

- **rectangle**

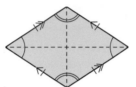

opposite sides equal and parallel; all angles equal (90°); diagonals equal

- **rhombus**

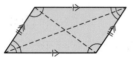

all sides equal; opposite angles equal; opposite sides parallel; diagonals *not* equal but cross at 90°

- **parallelogram**

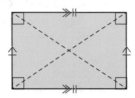

opposite sides equal and parallel; opposite angles equal; diagonals *not* equal

- **kite**

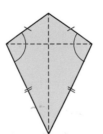

two pairs of adjacent equal sides; one pair of opposite equal angles; diagonals *not* equal but cross at 90°

- **trapezium**

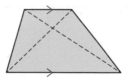

one pair of opposite parallel sides; diagonals only equal in an isosceles trapezium

There are some other special quadrilaterals.

- **delta (arrowhead) kite**   one angle is reflex (more than 180°)

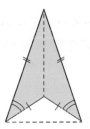

- **isosceles trapezium**   the two non-parallel sides equal; two pairs of adjacent angles equal; diagonals equal

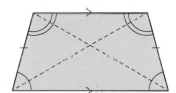

> **Key words:** square, rectangle, rhombus, parallelogram, kite, trapezium, delta, isosceles trapezium, parallel, diagonals, angles

Train

1 Copy the quadrilaterals below and draw their diagonals.   (9)

(a)   (b)   (c)

Write down underneath each quadrilateral all of the sentences below that are true for that shape.

The diagonals are equal.   One diagonal is a line of symmetry.

One diagonal bisects the other.   Both diagonals are lines of symmetry.

The diagonals meet at right angles.

Try

2 Which of these shapes is a quadrilateral?   (2)

rhombus   pentagon   hexagon   equilateral triangle

3 Draw a rhombus. Write down all the facts you know about its angles, sides, diagonals and lines of symmetry. Show them on the diagram, using appropriate marks.   (6)

Test

4 How many of the quadrilaterals below have diagonals of equal length?   (1)

square   rectangle   kite   rhombus   parallelogram   isosceles trapezium

5 All my sides are equal, all my angles are 90° and I have 4 lines of symmetry. What shape am I?   (1)

square

6 I have 2 pairs of equal parallel sides, all my angles are 90° and my diagonals are equal. What shape am I?   (1)

rectangle

7 I have no lines of symmetry but one pair of parallel sides. What shape am I?   (1)

H: Angles and lengths in quadrilaterals; drawing quadrilaterals

About this topic: In this section, you will continue working with quadrilaterals. You will revise using properties of quadrilaterals to make accurate drawings and to calculate unknown angles and side lengths. You will need a protractor, ruler and compasses.

Angle sum in quadrilaterals

The angle sum of a quadrilateral is 360°
Check by measuring the angles in these quadrilaterals with a protractor.

A square has 4 angles of 90°
The angle sum is 360°
(4 × 90° = 360°)

A kite has only one pair of equal angles. Measure the angles. The angle sum is 360°

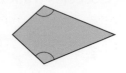

In this isosceles trapezium, a blue angle plus a red angle equals 180° because they are interior angles inside parallel lines.

There are two pairs, so the angle sum is 360°

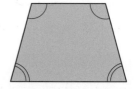

Finding the size of an unknown angle

You can use the angle sum of a quadrilateral and the properties of that quadrilateral to work out the size of an unknown angle.

Work out the size of the angles *B*, *C* and *D* in this parallelogram.

We can write angle *A* = 60° as ∠*A* = 60°

Opposite angles in a parallelogram are equal, so ∠*C* = 60°

∠*A* + ∠*B* = 180° → ∠*B* = 180° − 60° = 120°

Opposite angles are equal, so, ∠*D* = 120°

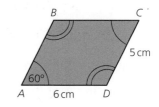

Finding unknown lengths

You can use what you have already learned about the properties of quadrilaterals to work out unknown lengths.

In parallelogram *ABCD* above:

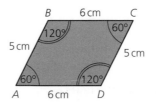

- *BC* is opposite to *AD*, so *BC* = 6 cm
- *AB* is opposite to *DC*, so *AB* = 5 cm

Drawing quadrilaterals

Draw this kite accurately.

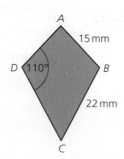

First, make a sketch. Calculate the unknown lengths and angles and write them on your sketch.

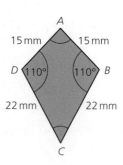

Follow these steps to draw the shape accurately.

Step 1 Draw the line *DC* = 22 mm

D ――――――― C

Step 2 Measure ∠*D* = 110° with your protractor. Draw a line from point *D*. Measure 15 mm along the line and mark point *A*.

Step 3 Use your compasses to make two arcs where *AB* (set at 15 mm) and *BC* (set at 22 mm) meet. Mark the intersection as point *B*.

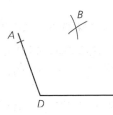

Step 4 Join the points and label known sides and angles.

Measure the unknown angles: ∠*A* = 87° and ∠*C* = 53°

Check the angle sum: 110° + 110° + 87° + 53° = 360°

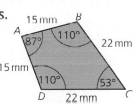

> **Key words:** angles, lengths

Train

1 Look at the square *ABCD*. (5)

(a) Work out the:

   (i) length *AB*   (ii) length *BC*   (iii) size of angle *D*.

(b) What is ∠*A* + ∠*B* + ∠*C* + ∠*D*?

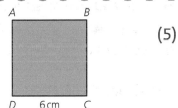
Not to scale

Try

2 Look at rectangle *ABCD*. (6)

(a) Work out the:

   (i) length *AD*   (ii) length *AD*   (iii) size of angle *A*?

(b) Draw *ABCD* accurately and measure the lengths of the diagonals *AC* and *BD*.

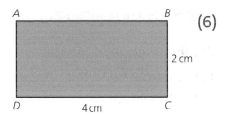

Test

3 Look at the right-angled trapezium *EFGH*. (6)

(a) Work out the size of these angles.

   (i) ∠*H*   (ii) ∠*E*   (iii) ∠*F*

(b) Draw *EFGH* accurately and measure the lengths *EF* and *FG*.

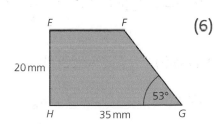

I: Polygons

About this topic: Polygons are two-dimensional shapes with many sides. Building on your revision of regular and irregular shapes and quadrilaterals, revise some other polygons.

Triangles are polygons with three sides.

Quadrilaterals are polygons with four sides.

Here are some other types of polygon.

● **pentagon**   5 sides

regular pentagon   irregular pentagon

● **hexagon**   6 sides

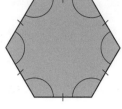
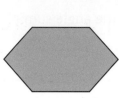

regular hexagon   irregular hexagon

● **heptagon**   7 sides
● **octagon**   8 sides
● **nonagon**   9 sides
● **decagon**   10 sides

A **regular polygon** has all sides equal and all angles equal. An **irregular polygon** has some unequal sides and some unequal angles.

Angles in a regular polygon

You already know that a regular polygon has equal sides and equal angles. You also know the angle sum of a triangle and a quadrilateral. You can use this information to work out the size of each **interior** angle.

● In a regular (equilateral) triangle, each interior angle is $180° \div 3 = 60°$
● In a regular quadrilateral (a square), each interior angle is $360° \div 4 = 90°$

You can use these formulae to work out the interior angles of the other regular quadrilaterals.

Exterior angle $= \dfrac{360°}{n}$

Interior angle $= 180° - \dfrac{360°}{n}$

where n is the number of sides in a regular polygon

The diagram shows that each pair of interior and exterior angles meet on a straight line and so add up to 180°

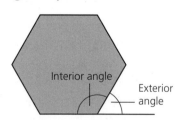

Interior angle

Exterior angle

Work out:

(i) the exterior angle and   **(ii)** interior angle of a regular 10-sided polygon.

(i) Exterior angle $= \dfrac{360°}{n}$, where $n = 10$ **(ii)** Interior angle $= 180° - 36° = 144°$

$= \dfrac{360°}{n} = 36°$

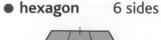

When you are asked to calculate the interior angle, you must start by calculating the exterior angle unless it is given.

Key words: regular polygon, irregular polygon, interior angle, exterior angle

Train

1 (a) Draw:
    (i) a regular pentagon   (ii) a regular hexagon. (2)
  (b) Write down the order of rotational symmetry of each shape. (2)
  (c) How many lines of symmetry does each shape have? (2)

2 Copy and complete this table. (6)

Number of sides, n	Name of regular polygon	Size of exterior angle	Size of interior angle
3			
4			

Try

3 Copy and complete this table. (15)

Number of sides, n	Name of regular polygon	Size of exterior angle	Size of interior angle
5			
6			
7			
8			
9			

Test

4 (a) How many lines of symmetry does a regular octagon have? (1)
  (b) What is the order of rotational symmetry of a regular octagon? (1)

5 Copy and complete these sentences.
  (a) The interior angle and the exterior angle of a regular polygon add up to ... (1)
  (b) The exterior angle of a regular polygon is calculated by the formula ... (1)
  (c) The interior angle of a regular polygon is equal to ... minus ... (1)

J: Circles

About this topic: In this section, you will revise circle terms and practise drawing circles accurately.

Circle terms

Look at this circle.

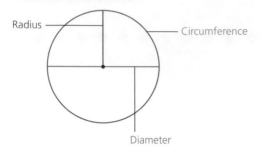

- The circumference is the boundary or perimeter of a circle.
- The radius is a straight line from the centre of the circle to a point on the circumference.
- The diameter is a straight line between two points on the circumference that passes through the centre of the circle.

Here are some other useful terms.

- An arc is part of the circumference of a circle.
- A sector is a slice between two radii and an arc.
- The angle at the centre is 360°

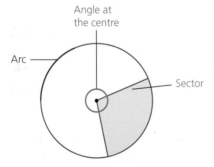

It is useful to remember that the angles at the centre of a circle add to 360° when you are drawing pie charts. You will revise pie charts in Chapter 6

Properties of circles

- The sum of angles at the centre of a circle is 360°
- A circle has an infinite number of lines of symmetry.
- A circle will always look the same when rotated.
- The distance from the centre of a circle to any point on its circumference (the radius) is always the same.

Drawing circles

The best way to draw a circle is with compasses. Extend the compasses to the length of the radius and put the point on your paper. Then draw the circle around the point.

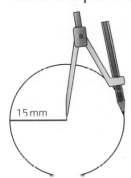

15 mm

> **Key words:** circle, circumference, radius, diameter, sector, arc

Train ●

1   If the radius of a circle is 4 cm, what is its diameter? (1)

2   If the diameter of a circle is 12 cm, what is its radius? (1)

3   If the diameter of a circle is 18 m, what is its radius? (1)

Try ●

4   **(a)** Draw a circle with a diameter of 4 cm. (1)

   (b) Mark the radius and the diameter with the correct lengths. (1)

   (c) Use a piece of string to measure the circumference of the circle. (1)

Test ●

5   What is the diameter of a circle with a radius of 3.5 cm? (1)

6   What is the radius of a circle with a diameter of 25 cm? (1)

7   **(a)** Draw a circle with a radius of 2 cm. (1)

   (b) Draw 2 lines of symmetry. (1)

   (c) Measure the angles at the centre of each sector and write them down. (1)

   (d) What is the angle sum at the centre of the circle? (1)

8   Draw a set of concentric circles, like the ones shown, with these measurements.

   (a) radius 2.5 cm (1)

   (b) radius 35 mm (1)

   (c) diameter 80 mm (1)

K: Perimeters of squares, rectangles and compound shapes

About this topic: In this section you will revise the methods for working out the perimeters of two-dimensional shapes, starting with squares, rectangles and compound shapes.

The **perimeter** of a shape is the distance around it. It can be measured in any unit of length, including millimetres (mm), centimetres (cm), metres (m) and inches.

Perimeter of a square

You know that the perimeter of a square of side 3 cm is
3 + 3 + 3 + 3 = 12 cm

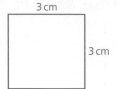

> When you calculate the perimeter of a shape, always check the units.

The perimeter of a square is the sum of the lengths of its four equal sides:
$b + b + b + b = 4b$

So the formula for the perimeter, P, of a square is: **$P = 4b$**

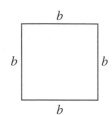

Calculate the perimeter of a square with side length 6 cm.

Perimeter = 4×6 cm = 24 cm

Perimeter of a rectangle

The perimeter of a rectangle of side 6 cm by 2 cm is 6 + 6 + 2 + 2 = 16 cm
You can also calculate this as 2(6 + 2) cm = 16 cm

The formula for the perimeter of a rectangle is:
$P = 2(l + w)$ where P is the perimeter, l is the length and w is the width, or
$P = 2(h + b)$ where, h is the height and b is the base of the rectangle

Work out the perimeter of a football pitch measuring 120 m by 90 m.

$P = 2(120 + 90)$ m = 2 × 210 m = 420 m

You can calculate the perimeter of other regular polygons in a similar way. The perimeter of a hexagon of side 2 cm is 12 cm
(6 sides, each of length 2 cm).

Perimeter of a compound shape

You can use what you know about the perimeter of rectangles and squares to work out the perimeter of a compound plane shape made from these shapes.

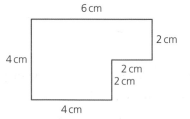

You may have to calculate unknown lengths before you can calculate the perimeter.

Perimeter = 6 + 2 + 2 + 2 + 4 + 4 = 20 cm

Work out the perimeter of this L-shaped room.
Calculate the missing lengths x and y first.

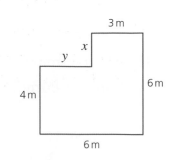

Missing lengths: $y = 6 - 3 = 3$ m and $x = 6 - 4 = 2$ m

Perimeter of the L-shaped room: 6 + 6 + 4 + 3 + 2 + 3 = 24 m

Key words: perimeter, plane shape, square, rectangle, compound shape

Train

1 Calculate the perimeter of each square. Remember to give your answers in the correct units. (3)

(a)

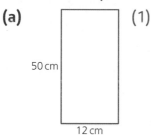

5.5 km

(b)

40 m

(c)

2 cm

Try

2 Calculate the perimeter of each rectangle. (1)

(a) (1)

50 cm

12 cm

(b) 6 m (1)

4 m

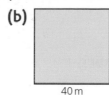

(c) 80 mm

30 mm

Test

3 Calculate the perimeters of these compound shapes. Work out any unknown lengths first. (3)

(a)

2 m

3 m

2 m

6 m

(b)

60 mm

20 mm

10 mm

40 mm

30 mm

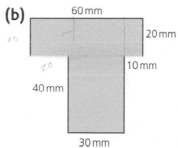

(c)

400 cm   8 m

400 cm

10 m

18 m

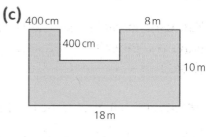

4 (a) An artist paints a picture on a canvas measuring 25 cm by 40 cm. What is the perimeter of the canvas? (1)

(b) The picture is then framed. The final measurements of the framed picture are 30 cm by 45 cm. What is the perimeter of the framed picture? (1)

(c) Write your answer to part (b) in metres? (1)

(d) The wood to frame the picture costs £4.50 per metre. How much will it cost to buy the exact length of wood to frame the picture? (1)

L: Areas of squares, rectangles and compound shapes

About this topic: Building on your work on perimeter, in this section you will revise how to calculate the areas of squares, rectangles and compound shapes.

The two-dimensional space inside a **plane shape** is called the **area**. Areas are usually calculated rather than measured. Units include square millimetres (mm^2), square centimetres (cm^2), square metres (m^2) and square kilometres (km^2).

Calculating the area of a square

You can work out the area of a square by counting centimetre squares inside the shape or by multiplying the number of rows of centimetre squares by the number of centimetre squares in each row: $2 \times 2 = 4\,cm^2$

In a square, the number of rows of squares will always be the same as the number of squares in each row because all the sides of a square are the same length.

The formula for calculating the area of a square is: $A = l^2$ where A is the area of the square and l is the side length.

The area of this square is $4\,cm^2$

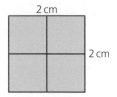

Calculating the area of a rectangle

You can use a similar method to calculate the area of a rectangle. The formula for the area of a rectangle is:

$A = l \times w$ where A is the area, l is the length and w is the width of the rectangle, or

$A = h \times b$ where A is the perimeter, h is height and b is base of the rectangle.

This rectangle is 2 rows of 6 squares. The area of the rectangle is $2 \times 6 = 12\,cm^2$

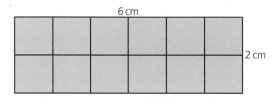

Calculating the area of a compound shape

You can use what you know about the area of rectangles and squares to work out the area of **compound shapes** made from these shapes.

You can either split the shape into two rectangles and add their areas or subtract the area of the 'missing' rectangle from the 'whole' large rectangle.

To work out the area of a regular hexagon, or any other shape, you can count centimetre squares inside it.

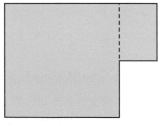
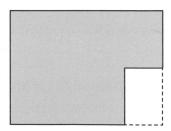

The area of each triangle below is $4\,cm^2$ (half the area of the rectangle in which it fits).

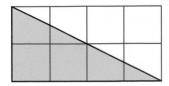
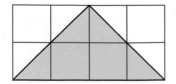
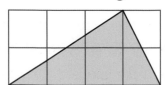

Work out the area of floor shown in this plan.

First, work out the unknown lengths. $x = 6 - 3 - 3\,m$ and $y = 7 - 3 = 4\,m$

Separate the floor plan into 2 rectangles, A and B. Then calculate the area of each rectangle.

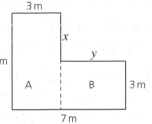

Area of rectangle A = $6 \times 3 = 18\,m^2$

Area of rectangle B = $4 \times 3 = 12\,m^2$

The combined area, A + B = $18\,m^2 + 12\,m^2 = 30\,m^2$

Always check the units given on a diagram. Check that you write the area with the correct units.

Key words: area, plane shape, square, rectangle, compound shape

Train

1 Calculate the area of each square. Remember to give your answers in the correct units. (3)

(a)
 49 cm

7 cm

(b) 25m

5 m

(c) 400 km.

20 km

Try

2 Calculate the area of each rectangle. (3)

(a) 6 mm
12 mm 72

(b) 150 cm
1500
50 cm
7500

(c) 30 m
750
25 m

Test

3 Calculate the area of each compound shape. Work out any unknown lengths first. (6)

(a)
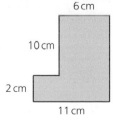
6 cm
10 cm
2 cm
11 cm

(b)
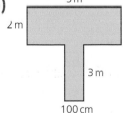
5 m
2 m
3 m
100 cm

(c)
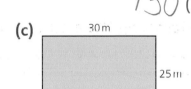
65 cm   65 cm
100 cm
80 cm   120 cm
210 cm

22
22
‾‾
44

4 A square garden has side length 22 m.

(a) What is the area of the garden? 484 m   484 (1)
440
484

(b) What length of fencing is needed to enclose the garden completely? (1)

(c) Fence panels are 2 m wide. How many panels are needed to enclose the garden? (1)

22
22

M: Areas of triangles and parallelograms

About this topic: In this section you will extend your revision of calculating areas to triangles and parallelograms. Thinking about how different shapes are related to each other will help you master this topic.

Calculating the area of a right-angled triangle

You can think of a right-angled triangle as half a rectangle. Therefore, the area of a right-angled triangle is half the area of a rectangle with the same length (base) and width (height).

The formula for calculating the area of a rectangle is:

$A = l \times w$ where A is the area, l is the length and w is the width of the rectangle, or

$A = h \times b$ where A is the area, h is height and b is base of the rectangle.

So, the formula for calculating the area of a right-angled triangle is: $A = \frac{1}{2}(l \times w)$

Work out the area of this right-angled triangle.

Area $= \frac{1}{2}(l \times w) = \frac{1}{2}(7 \times 4) = 14\,cm^2$

Remember to check you have used the correct units.

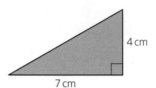
4 cm

7 cm

Calculating the area of a parallelogram

The diagram shows the relationship between the area of a parallelogram and the area of a rectangle with the same length and width.

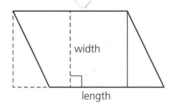
width

length

Note from the diagram that the measurements needed for l and w must be at right angles to each other.

The formula for calculating the area of a parallelogram is the same as the formula for calculating the area of a rectangle.

The formula for calculating the area of a parallelogram is: $A = l \times w$

Work out the area of this parallelogram.

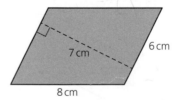
7 cm   6 cm

8 cm

Area of a parallelogram $= l \times w = 6 \times 7$
$= 42\,cm^2$

Calculating the area of any triangle

You can think of a non-right-angled triangle as half a parallelogram.

This means that the formula to calculate the area of any triangle is:
$A = \frac{1}{2}(l \times w)$

As with the formula for the parallelogram, the measurements needed for l and w must be at right angles to each other.

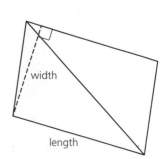
width

length

To measure the width of an obtuse-angled triangle, extend the base and draw a perpendicular to the top vertex of the triangle.

Calculate the area of this triangle.

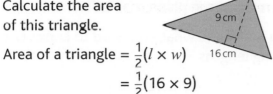

Area of a triangle = $\frac{1}{2}(l \times w)$

$= \frac{1}{2}(16 \times 9)$

$= 72 \text{ cm}^2$

Calculate the area of this obtuse-angled triangle.

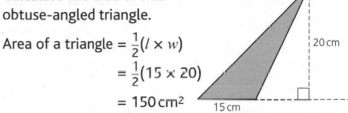

Area of a triangle = $\frac{1}{2}(l \times w)$

$= \frac{1}{2}(15 \times 20)$

$= 150 \text{ cm}^2$

Key words: right-angled triangle, parallelogram, triangle, area

Train

1 Work out the area of each right-angled triangle. (3)

(a)

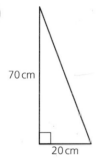

70 cm

20 cm

(b)

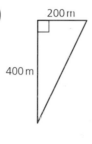

200 m

400 m

(c)

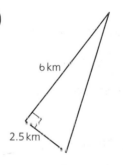

6 km

2.5 km

Try

2 Work out the area of each parallelogram. (3)

(a)

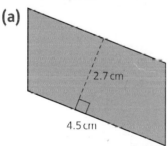

2.7 cm

4.5 cm

(b)

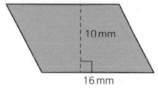

10 mm

16 mm

(c)

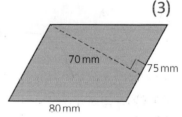

70 mm

75 mm

80 mm

Test

3 Work out the areas of these shapes. (3)

(a)

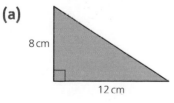

8 cm

12 cm

(b)

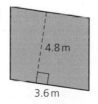

4.8 m

3.6 m

(c)

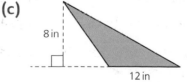

8 in

12 in

4 In triangle *ABC*, the base *AB* = 36 mm and the height is 4 cm. Work out the area of the triangle. (1)

5 In triangle *DEF*, the base *DE* = 4 cm and the height is 3.5 cm. Work out the area of the triangle. (1)

N: Calculating unknown lengths

About this topic: In this section you will practise using the area and perimeter formulae you have revised in the last few sections to work out unknown lengths in plane shapes.

Drawing a **sketch** is a good way of working out what you know and what you need to know. Use the **formulae** relating to the shape to work out any missing lengths.

P represents perimeter, A the area, b the base and h the perpendicular height in these formulae.

Perimeter of a square	$P = 4b$	Area of a square	$A = b^2$
Perimeter of a rectangle	$P = 2(b + h)$	Area of a rectangle	$A = b \times h$
Area of any triangle	$A = \frac{1}{2}(b \times h)$	Area of a parallelogram	$A = b \times h$

Work through these examples.

A rectangle has an area of 50 cm² and a height of 4 cm. How long is the base?

Draw a sketch:

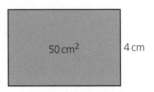

Area of a rectangle = $b \times h$	Write down the formula you need.
$50 = b \times 4$	Substitute what you know.
$b = 50 \div 4 = 12.5$ cm	Calculate.

The base of the rectangle is 12.5 cm

> **Always check that you have used the correct units.**

A parallelogram has an area of 48 km² and a base of 6 km. What is the perpendicular height?

Draw a sketch:

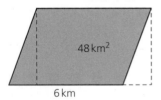

Area of a parallelogram = $b \times h$	Write down the formula you need.
$48 = 6 \times h$	Substitute what you know.
$h = 48 \div 6 = 8$ km	Calculate.

The perpendicular height of the parallelogram is 8 km

A triangle has an area of 21 cm² and a base of 6 cm. Work out the perpendicular height.

Draw a sketch:

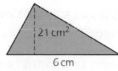

Area of a triangle $= \frac{1}{2}(b \times h)$ Write down the formula you need.

$21 = \frac{1}{2}(6 \times h)$ Substitute what you know.

$21 = 3 \times h$ so $h = 21 \div 3 = 7$ cm   Calculate.

The perpendicular height of the triangle is 7 cm

Key words: formula, substitute

Train

1 Work out the length of the base (b) in each shape. (6)

(a)

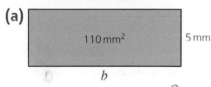

(b) $5\overline{)110}$ (c)

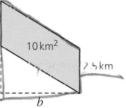

$5\overline{)110}$

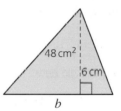

Try

2 A square area of lawn is 36 m². What is the length of each side of the lawn? (2)

3 A triangle has an area of 200 m² and a perpendicular height of 8 m. How long is the base? (2)

4 A parallelogram has an area of 264 cm² and a perpendicular height of 24 cm. How long is the base? (2)

Test

5 Work out the perpendicular height of this parallelogram. (2)

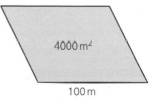

6 The school is planning build a tennis court, a netball court and a volleyball court.

(a) Work out the length of the tennis court if it has an area of 264 m² and a width of 11 m. (2)

(b) Work out the length of the netball court if it has an area of 450 m² and a width of 15 m. (2)

(c) Work out the area of the badminton court if it is 6 m wide and 13 m long. (2)

7 A triangle has an area of 144 cm² and a base of 12 cm. Work out its perpendicular height. (2)

8 A garage floor is rectangular. Two of its parallel sides each measure 5 m. The area of the floor is 16.5 m². What is the measurement of each of the other two sides? (2)

O: Three-dimensional shapes

About this topic: In this section, you will begin to revise three-dimensional (solid) shapes, starting with properties of some common solids and their nets.

Three-dimensional (3D) shapes have one more dimension than two-dimensional shapes. As well as length and width, they also have height. They have faces, edges and vertices (corners).

They may have:

● one or more **planes of symmetry**
● **axes of rotational symmetry**.

Properties of common 3D shapes

● **cube**   6 square faces; 8 vertices, 12 edges, 3 pairs parallel faces; every face perpendicular (at a right angle) to another face; 6 congruent square faces

● **cuboid**   6 rectangular faces; 8 vertices; 12 edges; 3 pairs parallel faces; every face perpendicular (at a right angle) to another face; 3 pairs of congruent rectangular faces

● **tetrahedron**   4 equilateral triangular faces; 4 vertices; 6 edges; 4 congruent triangular faces

● **square-based pyramid**   1 square face and 4 isosceles triangular faces; 5 vertices; 8 edges

● **triangular prism**   2 triangular end faces; 3 rectangular faces; 6 vertices; 9 edges

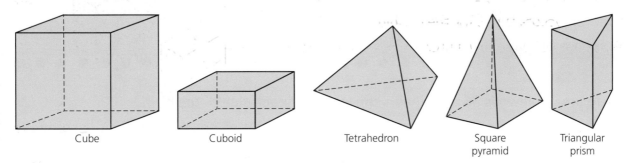

Cube   Cuboid   Tetrahedron   Square pyramid   Triangular prism

Nets

A **net** of a 3D shape:

● shows all the faces of the shape, with appropriate faces joined together
● can be folded up to make the shape.

The diagram shows one possible net for each of the 3D shapes shown above. (It is often possible to draw several different nets for one 3D shape.)

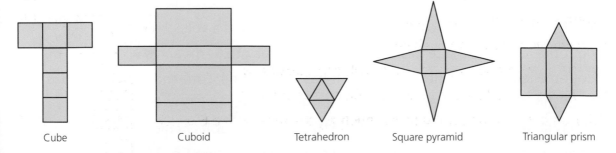

Cube   Cuboid   Tetrahedron   Square pyramid   Triangular prism

Look carefully at each net and work out how it can be folded to make the 3D shape. Look at which sides join to make each edge of the solid.

Key words: three-dimensional, 3D, cube, cuboid, tetrahedron, square pyramid, triangular prism, net

Train

1   These drawings show cuboids made from centimetre cubes.   (3)

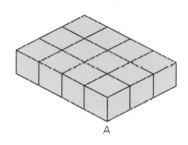 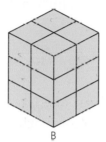

How many cubes are needed to build:

(a) cuboid A   (b) cuboid B?

Try

2   The diagram shows a cuboid made using centimetre-squared paper.   (6)

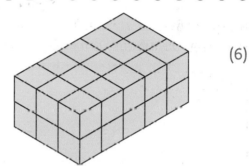

(a) Draw an accurate net for the cuboid.

(b) What area of paper is needed to make the cuboid?

Test

3   (a) Copy and complete this table. Do not look at the information on the previous page!   (12)

Solid shape	Number of faces	Number of vertices	Number of edges
Cube			
Cuboid			
Square pyramid			
Triangular prism			

(b) Draw three different nets of a cube.   (3)

4   This diagram shows the net of a die.   (4)

(a) How many faces does the die have?

(b) When the die is made, which number will be opposite 5?

(c) When the die is made, which number will be opposite 1?

(d) Which numbers will be on the faces touching the face with a 6 on it?

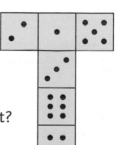

P: Capacity and volume

About this topic: You will continue working with three-dimensional shapes, looking at capacity and volume and revising the formula to calculate the volume of a cuboid.

Units of volume and capacity

Units of volume and capacity include:

Metric units   cubic centimetres (cm^3), cubic metres (m^3), millilitres (ml), litres (l)

Imperial units   pints, gallons

Here are some useful **conversions**:

$1000\,cm^3 = 1$ litre

$1\,000\,000\,cm^3 = 1\,m^3$ $(1\,m^3 = 100 \times 100 \times 100\,cm^3)$

$1000\,ml = 1$ litre

4.5 litres ≈ 1 gallon

Capacity usually refers to the amount that something, for example a bucket, could hold.

Volume is a measure of the three-dimensional space that something takes up.

The capacity of this container is 1 litre.
The volume of water inside the container is $\frac{1}{2}$ litre (0.5 l).

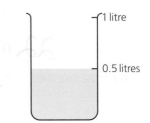

Calculating the volume of a cuboid

The diagram shows a cuboid measuring $4\,cm \times 3\,cm \times 2\,cm$ made from centimetre cubes.

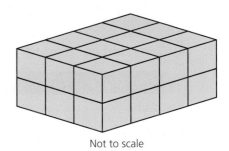

Not to scale

There are $4 \times 3 = 12$ centimetre cubes in each layer.
There are 2 layers, so there are $12 \times 2 = 24$ centimetre cubes altogether. The volume of the cube is $24\,cm^3$

The formula for calculating the volume of a cuboid is: volume = length × width × height

$V = l \times w \times h$

A cuboid has a length of 5 cm, width 4 cm and height 3 cm.

Calculate the volume of the cuboid.

$V = l \times w \times h$	Write down the formula.
$V = 5 \times 4 \times 3$	Substitute the values.
$= 60\,cm^3$	
The volume of the cuboid is $60\,cm^3$	Remember to give your answer in the correct units.

Using the volume formula to calculate unknown lengths

You can use the volume formula to calculate an unknown length in a cuboid in the same way you used area formulae to calculate unknown lengths.

A cuboid has length 8 cm, width 5 cm and volume 80 cm³. Work out the height of the cuboid.

$V = l \times w \times h$ Write down the formula.

$80 = 8 \times 5 \times h$ Substitute the values.

$80 = 40 \times h$

$h = 80 \div 40 = 2$

The height of the cuboid is 2 cm

> **Key words:**
> capacity,
> volume, cubic
> centimetres,
> cubic metres,
> millilitres, litres

Train

1 Work out the volume of cuboids with these dimensions.   (2)
   (a) 2 cm by 4 cm by 5 cm   **(b)** 10 m by 9 m by 3 m
2 A hollow cube has edge length 7 cm. What is the capacity of the cube?   (1)
3 How many centimetre cubes will fit into a box measuring 10 cm by 7 cm by 6 cm?   (1)

Try

4 The three models in the picture are made from centimetre cubes.   (6)

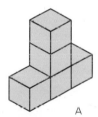 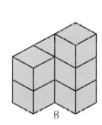 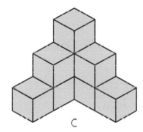

A   B   C

   (a) What is the volume of model A?   **(c)** What is the volume of model C?
   (b) What is the volume of model B?
5 Which has the greater volume, a cube with sides 6 cm or a cuboid measuring 4 cm by 4 cm by 12 cm?   (2)

Test

6 A swimming pool in the shape of a cuboid has length 20 m, width 10 m and depth 5 m. How much water could it hold?   (2)
7 Work out the heights of cuboids with these measures.   (3)
   (a) length 6 cm, width 10 cm, volume 120 cm³   **(c)** length 8 m, width 2.5 m, volume 20 m³
   (b) base area 36 cm², volume 180 cm³
8 Ella is designing a box to hold 12 cubes that each have edge length 1 cm. Her box is 1 cm × 1 cm × 12 cm. Write down the dimensions of all the other boxes that will hold exactly 12 cubes.   (4)

Q: Angles

In this section you will revise the properties of angles and how to measure them.

An angle is the amount of turn between two straight lines that meet. An angle can be any size from 0° (no turn) to 360° (one complete turn).

Angles $x + x + y + y = 360°$

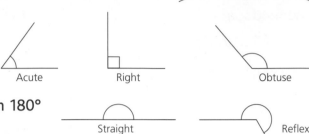

The two angles labelled x are equal in size, as are the two angles labelled y.

Types of angle

We classify angles by the size of the turn.

- An **acute angle** is less than 90°
- A **right angle** is 90°
- An **obtuse angle** is greater than 90° but less than 180°
- A straight line has an angle of 180°
- A **reflex angle** is greater than 180° but less than 360°

Acute   Right   Obtuse

Straight   Reflex

Measuring angles

You can use a protractor to measure angles.

Follow the steps below to make sure you position your protractor correctly.

- Place the centre mark of the protractor where the two arms of the angle meet.
- Check that the zero line is placed on one arm of the angle.
- Count round from zero until you reach the second arm of the angle.
- Read off the angle on the scale.

Protractor

Read here

Zero line

You will need to add 180° for a reflex angle unless you have a circular (360°) protractor.

Drawing angles

Follow these steps to draw an angle.

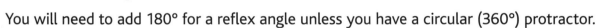
Use a sharp pencil to draw angles.

- Draw the first arm of the angle.
- Put the protractor with the zero line on the first arm, with the centre mark of the protractor on the end of the line.
- Count round from zero until you reach the required angle and mark a small line or dot.
- Draw a line from the centre mark to your new mark.
- Label your angle.

Estimating angles

Some questions may ask you to estimate an angle. This example shows you a useful method.

Estimate the size of this angle in degrees.

Ask yourself: What type of angle is it?   Acute

Is more or less than half a right angle (45°)?   More

Is it nearer to 45° or 90°?   Nearer 90°

Make a guess: About 70°

Key words: angle, acute, right, obtuse, straight, reflex, protractor

Train

1 For each angle below: (6)

  (i) write down the type of angle   (ii) *estimate* the size of the angle.

(a) 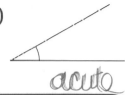 45° acute

(b)  115° obtuse

(c)  190° reflex

Try

2

(a) What type of angle is this? (1)

(b) Measure the angle. (You may need to extend the lines on a copy of the diagram.) (1)

3 Draw an angle of exactly 75° (1)

4 Estimate the size of this angle. (1) 5 Calculate the value of x below. (1)

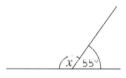 x 55°

Test

6 Write down the letter of an angle inside this shape which is: (4)

  (a) an obtuse angle   (c) a reflex angle

  (b) an acute angle   (d) a right angle.

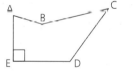

7 Calculate the size of each lettered angle. Do not measure them. (2)

(a) 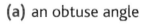 110° x

(b) 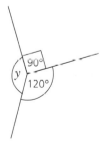 90° y 120°

8 (a) Draw an angle of 50°. What type of angle is this? (2)

  (b) Draw an angle of 155°. What type of angle is this? (2)

  (c) Draw an angle of 220°. What type of angle is this? (2)

S: Angle facts and calculating angles

About this topic: You already know from your revision that you can use properties of shapes and angle facts to calculate unknown angles. In this section you will build on this and revise and practise using some more useful angle facts.

These **pairs of angles** are all **equal**.

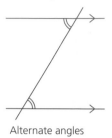

Alternate angles

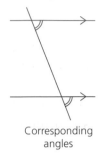

Corresponding angles

Vertically opposite angles

These sets of **angles add up to 180°**

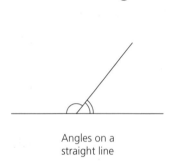

Angles on a straight line

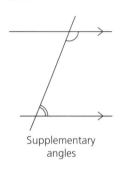

Supplementary angles

Angles in a triangle

You can use these angle facts to calculate unknown angles.

What is the value of x?

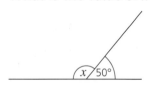

$x = 180° - 50° = 130°$
Angles on a straight line add up to 180°

What is the value of z?

$z + z = 180° - 60° = 120°$ Angles in a triangle add up to 180°

$z = 120° \div 2 = 60°$

These sets of **angles add up to 360°**

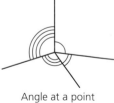

Angle at a point

Angles in a quadrilateral

What is the value of y?

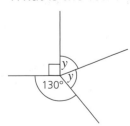

$y + y + 130° + 90° = 360°$ Angles at a point add up to 360°

$y + y = 360° - 130° - 90° = 140°$

$y = 70°$

Work out the values of x, y and z

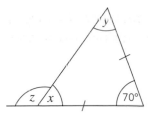

$x + y + 70° = 180°$, so $x + y = 110°$ Angles in a triangle add up to 180°

Isosceles triangle (2 equal sides marked) so $x = y \rightarrow x = 55°$ and $y = 55°$

$z + x = 180°$ so $z = 180° - 55° = 125°$ Angles on a straight line equal 180°

So, $x = 55°$, $y = 55°$ and $z = 125°$

Key words: alternate angles, corresponding angles, vertically opposite angles, angles on a straight line, supplementary angles, angles at a point, angles in a quadrilateral

Train

1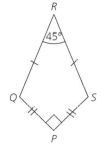

(a) What is the name given to this quadrilateral?   (1)

A: square   B: trapezium   C: rectangle   D: rhombus   E: kite

(b) What type of angle is PQR?   (1)

A: acute   B: right   C: obtuse   D: straight   E: reflex

(c) What is the size of angle PQR?   (1)

A: 100°   B: 112.5°   C: 110°   D: 135°   E: 225°

2   Work out the values of the unknown angles.   (9)

(a)
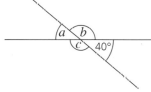

(b)
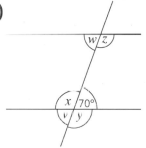

(c)
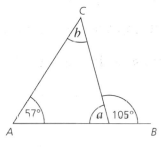

Try

3   Use angle facts to work out the unknown angles. *Do not measure.*   (4)

(a)
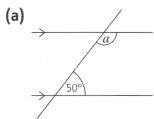

(b)
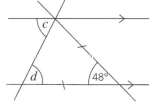

(c)
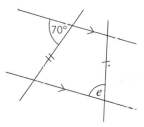

Test

4   Calculate the sizes of the unknown angles. Write down the angle facts that relate to each calculation.   (11)

(a)
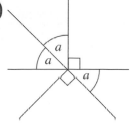

(b)
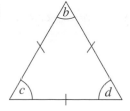

(c)
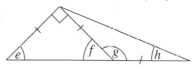

T: The eight-point compass

About this topic: In this topic you will revise compass points, compass bearings and scale drawings of position.

Compass points

The four main compass points are north (N), south (S), west (W) and east (E).

To help you remember, W is on the left and E is on the right – W E forms the word 'WE'.

There are four more points between the four main points: north-east (NE), south-east (SE), south-west (SW) and north-west (NW).

Note, the letter N or S is always placed first in these four in-between points.

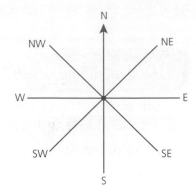

In this scale drawing:
- Archie is north of Ed
- Ed is west of Oliver
- Ed is north-east of Angus
- Angus is south-west of Ed.

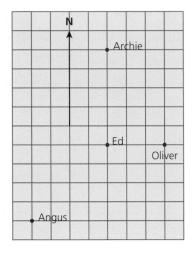

Bearings

You can use the eight-point compass to describe a direction. However, three-figure bearings allow us to describe direction more accurately.

Each of the eight points has a bearing.

N	000°	S	180°
NE	045°	SW	225°
E	090°	W	270°
SE	135°	NW	315°

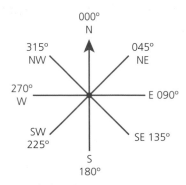

Bearings are always given as measurements clockwise from the north and as three figures.

Scale drawings

In the scale drawing in the example, 0.5 cm represents 2 m.

Archie is 10 m north of Ed (5 squares).

Ed is 6 m west of Oliver (3 squares).

Ed is approximately 11 m north-east of Angus.

Key words: compass points, bearings

Train

1 The scale drawing shows five pupils standing on a patio made with square concrete slabs of side length 1 m.

Write down the compass direction and bearing of: (10)

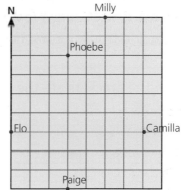

(a) Camilla from Flo

(b) Phoebe from Camilla

(c) Milly from Phoebe

(d) Flo from Paige

(e) Phoebe from Paige.

Try

2 (a) Draw an eight-point compass and label the directions and bearings. (4)

(b) I face north-east when going towards the shops from my front door.
I then turn 90° clockwise. In which direction am I now facing? (1)

Test

3 In the scale drawing in question 1, 0.5 cm represents 1 m. Write down the distance, in metres, between each pair of pupils listed in parts (a)–(e). (5)

4 On a copy of the diagram below, draw lines to show the path taken by following these instructions. Start at the dot. (7)

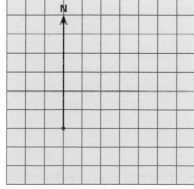

(a) Go 1.5 cm N

(b) Go 1 cm W

(c) Go 1 cm N

(d) Go 4 cm E

(e) Go 2.5 cm S

(f) Go 2 cm NW

(g) Go 2 cm SW. Draw and label the final position X

U: Reflection

About this topic: In the final three sections in this chapter, starting with reflection, you will revise the three transformations that move a shape without changing its shape or size.

A shape can be **reflected** in a line to form a mirror **image** of the shape. The original shape (**object**) and its **image** are identical even though they are different ways round. This means the object and its image are **congruent**.

The reflection is the image of the original shape in a **line of symmetry**.

Reflections on a grid

Transformations are often shown on co-ordinate grids. The co-ordinate grid below has vertical and horizontal axes, with values ⁻6 to 6

The numbers in the brackets are the **co-ordinates** of the points.

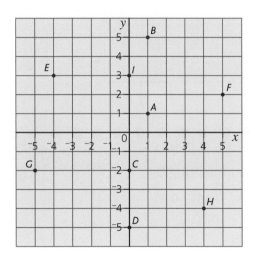

$A(1, 1)$	$B(1, 5)$	$C(0, ^-2)$
$D(0, ^-5)$	$E(^-4, 3)$	$F(5, 2)$
$G(^-5, ^-2)$	$H(4, ^-4)$	$I(0, 3)$

Remember, in a pair of co-ordinates, the first number is the x co-ordinate and the second number is the y co-ordinate.

An object and its image are the same perpendicular distance from the mirror line.

This co-ordinate grid shows two examples of reflection on a grid.

- Triangle B is the image of triangle A when it is reflected in the y-axis.
  The point X on triangle A is 3 grid squares to the right of the y-axis.
- The reflection of this point (X') is 3 grid squares to the left of the y-axis.
- Triangle C is the image of triangle A when it is reflected in the x-axis.

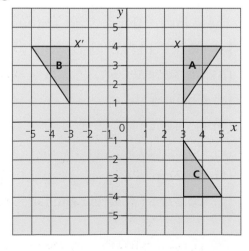

Key words: transformation, reflection, object, image, congruent, line of symmetry, axes, co-ordinates

Train

1 Copy the diagram and reflect the shape in the mirror line **m**. (3)

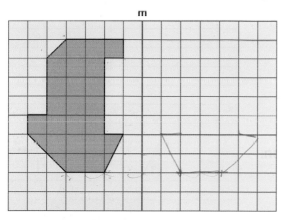

Try

2 Copy this grid and triangle A.

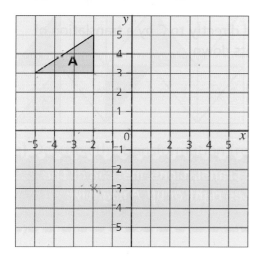

(a) Reflect triangle A in the x-axis. Label the image X. (2)

(b) Reflect triangle A in the y-axis. Label the image Y. (2)

Test

3 (a) Draw a co-ordinate grid like the one in question 2. Draw a shape with vertices at (1, 1), (1, 3), (2, 4), (4, 4), (5, 3) and (5, 1). Label it A. (3)

(b) Draw a reflection shape A in the x-axis and label it X. Write down the co-ordinates of the vertices. (4)

(c) Draw a reflection of shape A in the y-axis and label it Y. Write down the co-ordinates of the vertices. (4)

W: Rotation of a shape on a grid

About this topic: In this section you will revise **rotation**. Rotation is the third and final translation that can change the position of a shape without changing its shape or size.

A shape can be **rotated** through an angle **about a point**.
The most common angles are 90° and 180°

The original shape (**object**) and its **image** are identical even though the image has been turned. The object and its image are **congruent**.

To rotate an image you need three pieces of information:

- the centre of rotation
- the angle of rotation
- the direction of rotation, clockwise or anticlockwise, unless it is 180°

> Use a piece of tracing paper to help you with rotations. Trace the original shape, put the tip of pencil on the point of rotation and rotate the required number of degrees clockwise or anticlockwise.

Image of A after rotation through 180° around point P

Centre of rotation

Image of A after rotation through 90° anticlockwise around point P

Use tracing paper to check these rotations.

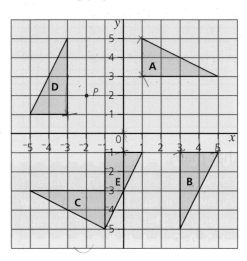

B is the image of A after a rotation of 90° clockwise about the origin.

C is the image of A after a rotation of 180° about the origin.

D is the image of A after a rotation of 90° anticlockwise about the origin.

E is the image of A after a rotation of 90° clockwise about the point P (⁻2, 2)

> **Key words:** rotation, point, clockwise, anticlockwise

Train

1 Write out in full the rotation that maps: (4)

   (a) A to B

   (b) A to C

   (c) A to D

   (d) D to E.

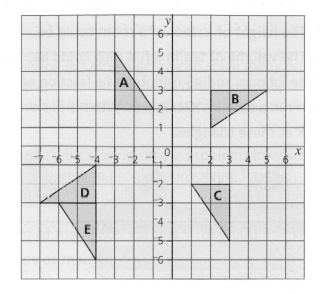

Try

2 Copy this grid and shape A. (6)

   (a) Draw the image of A after a rotation of 90° anticlockwise about the origin. Label it B.

   (b) Draw the image of A after a rotation of 180° about the point (⁻1, ⁻1). Label it C.

   (c) Draw the image of A after a rotation of 180° about the point (3, 0). Label it D.

Test

3 Draw a grid like the one in question 2 (8)

   (a) Draw triangle A with vertices at (⁻3, 1), (⁻5, 1) and (⁻4, 3)

   (b) Draw the image of A after the rotation of 90° clockwise about the origin. Label it B.

   (c) Draw the image of A after the rotation of 180° about the origin. Label it C.

   (d) Draw the image of C after the rotation of 90° anticlockwise about the point (4, ⁻3). Label it D.

A: Missing numbers; missing operations

About this topic: The next step in your mathematical journey is to **use** some of the many mathematical rules you have learned, to solve problems. In this section, you will revise how to solve puzzles that involve missing numbers or operations. This is simple algebra.

In situations where you know some information but some other information is missing, you can use what you know to work out the unknowns.

Missing numbers

Work through these two examples. Some of the numbers are missing, but you should be able to use the familiar layout of the calculations to help you work out what they are.

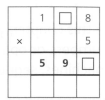

Missing operations

In these examples, some of the operations (addition, subtraction, multiplication and division) are missing.

$8 \boxed{} 5 = 13$ 13 is larger than 5 and 8, so the missing operation must be + or ×

$8 + 5 = 13$

$5 \boxed{} 4 = 10 \boxed{} 2$ Knowing your times tables really helps.

$5 \times 4 = 10 \times 2$

$(2 \boxed{} 3) \boxed{} 6 = 30$

$(2 + 3) \times 6 = 30$

$5 \times 6 = 30$

> Remember to follow the BIDMAS rules. You are unlikely to get the correct answer if you don't do calculations in the correct order.

Problem solving

In the examples above, you had to find missing numbers or operations. Some questions will only give you an answer and you will need to write your own simple statement before you can work out the missing pieces of information. You can use a letter or a symbol to represent an unknown number.

The total cost of 4 cinema tickets is £20. What is the cost of 1 ticket?

You could write this as: ■ + ■ + ■ + ■ = £20   where ■ is the cost of 1 ticket

or as: 4 × ■ = £20

■ = 20 ÷ 4 = 5   The price of each ticket is £5

You can extend this idea to solve a group of questions. Finding the answer to one may help you solve the others.

∗ + ∗ = 8   and   ∗ × ▲ = 8   Find the values of ∗ and ▲

∗ = 8 ÷ 2 = 4   Start by solving the statement with one unknown number.

4 × ▲ = 8   Substitute ∗ = 4 into the second equation to find the value of ▲

▲ = 8 ÷ 4 = 2   So, ∗ = 4 and therefore ▲ = 2

Train

1 Work out the missing numbers. (3)

(a)

	☐	3	8	☐
+		1	☐	6
	1	5	3	6

(b)

	7	5	☐
−		4	2
	☐	☐	6

(c)

	1	5	8
×			☐
	☐	1	6

2 Work out the missing operations (+, −, ×, ÷). (3)

(a) 34 ☐ 4 = 30   **(b)** 19 = 26 ☐ 7   **(c)** 12 ☐ 6 ☐ 4 = 8

Try

3 Solve each of these. (3)

(a)

	1	☐	6	☐
×				5
	5	3	1	0

(b)

	3	☐	9
2	☐	7	8

(c)

		2	4	3
×			2	☐
		2	4	3
+	4	☐	☐	☐
	☐	☐	☐	☐

4 Solve these by working out the missing operations (+, −, ×, ÷). (3)

(a) 30 ☐ 3 = 10   **(b)** 4 ☐ 3 = 8 ☐ 4   **(c)** (12 ☐ 2) ☐ 6 = 9 ☐ 2

Test

5 Solve these. (2)

(a) 8 ☐ 7 ☐ 4 = 11

(b)

	☐	☐	6
−	1	5	☐
	1	0	2

6 Write down the value of both symbols in each set of statements.

(a) ✚ × ✚ × ✚ = 27

◆ − ✚ = 12 (2)

(b) ● + ● + ● = 15   ● × ◆ = 20   ● − ◆ = 1 (3)

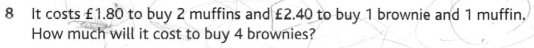

7 The sum of two numbers is 8. The difference between the two numbers is 4. What are the numbers? (2)

8 It costs £1.80 to buy 2 muffins and £2.40 to buy 1 brownie and 1 muffin. How much will it cost to buy 4 brownies? (2)

B: Function machines

About this topic: In this section you will look at look at how rules or functions can be applied to one number to change it into a different number.

A **function machine** changes one number (an **input**) into another number (an **output**) by applying one or more functions to the input number.

This function machine consists of the single function 'add 5'

input				output
1	→		→	6
7	→		→	12
23	→	+ 5	→	28
⁻3	→		→	2
0	→		→	5

This function machine consists of two functions 'multiply by 2' and 'add 3'

input	× 2		+ 3	output
1	→	2	→	5
7	→	14	→	17
23	→	46	→	49
⁻3	→	⁻6	→	⁻3
0	→	0	→	3

> When a function machine has two or more functions, it is a good idea to write down the result after each operation.

The functions must be carried out in the order in which they are written. The outputs will not be the same if the operations are applied in the wrong order. Look at what happens if we reverse the functions in the above example.

input	+ 3		× 2	output
1	→	4	→	8
7	→	10	→	20
23	→	26	→	52
⁻3	→	0	→	0
0	→	3	→	6

The pair of numbers (input and output) is called an **ordered pair**. The input number always comes first.

The ordered pairs for the function machine above are (1, 8), (7, 20), (23, 52), (⁻3, 0) and (0, 6) This is called a sequence of ordered pairs.

> **Key words:** function machine, input, output, ordered pair

Train

1 Each letter represents an unknown input or output of the function machine.
   Work out the value of each letter.   (6)

   (a) input   output   **(b)** input   output

   2 → ┌─────┐ → *a*   ⁻3 → ┌─────┐ → *a*
   0 → │ + 5 │ → *b*   *b* → │ − 3 │ → 2
   c → └─────┘ → 6   *c* → └─────┘ → 7

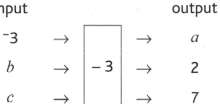

Try

2 Each letter represents an unknown input, output or intermediate value of the
   function machine. Work out the value of each letter.   (11)

   (a)
   a → ┌─────┐ → 5 → ┌─────┐ → *b*
   4 → │ + 2 │ → *c* → │ × 2 │ → *d*
   e → └─────┘ → 1 → └─────┘ → 2

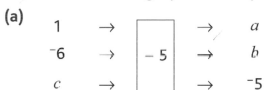

   (b)
   5 → ┌─────┐ → *a* → ┌─────┐ → *b*
   ⁻5 → │ × 2 │ → *c* → │ ÷ 5 │ → *d*
   e → └─────┘ → 40 → └─────┘ → *f*

Test

3 Calculate the missing input and output values in these function machines.   (9)

   (a)
   1 → ┌─────┐ → *a*
   ⁻6 → │ − 5 │ → *b*
   c → └─────┘ → ⁻5

   (b)
   10 → ┌─────┐ → *a* → ┌─────┐ → *b*
   ⁻6 → │ + 2 │ → *c* → │ ÷ 4 │ → *d*
   e → └─────┘ → 40 → └─────┘ → *f*

4 Work out the functions A, B and C and the missing inputs and outputs in these
   function machines.   (7)

   (a)
   12 → ┌─────┐ → 9
   0 → │ A │ → *a*
   b → └─────┘ → 3

   (b)
   14 → ┌─────┐ → 18 → ┌─────┐ → *a*
   8 → │ B │ → 12 → │ C │ → 2
   b → └─────┘ → ⁻6 → └─────┘ → ⁻1

C: Sequences and number puzzles

About this topic: You need to be able to spot patterns or rules in sequences in both mathematics and non-verbal reasoning. You will also need to play detective with numbers in verbal reasoning. Mastering the work in this topic will help you with all these challenges.

Sequences

A **sequence** is a list of numbers, shapes or patterns that follow a **rule**.

The individual numbers in a number sequence are the **terms** of the sequence.

In the sequence 1, 5, 9, 13, 17, ...   the rule for finding the next term is 'add 4'

In the last section you revised functions, which are rules that change one number into another number. You can use a function machine to produce a sequence by applying the rule (function) over and over again.

$$1 \rightarrow \boxed{+4} \rightarrow 5$$
$$5 \rightarrow \boxed{+4} \rightarrow 9$$
$$9 \rightarrow \boxed{+4} \rightarrow \text{... and so on.}$$

You can also write the sequence like this:

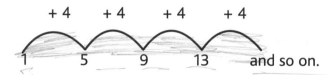

and so on.

Rules for sequences can involve 'counting on', 'counting back', multiplying, dividing and more.

100, 95, 90, 85, 80, ...	The rule is 'subtract 5'
1, 3, 9, 27, 81, ...	The rule is 'multiply by 3'
40, 20, 10, 5, ...	The rule is 'divide by 2'
1, 2, 3, 5, 8, ...	The rule is 'add the last two terms'.

Some rules are easier to spot than others. Working out the difference between consecutive terms will help you to spot the rule.

What are the next two terms in this sequence?   0, 3, 8, 15, ..., ...

Write out the sequence with the differences.

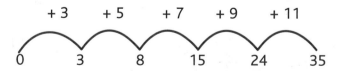

The differences in this sequence are the sequence of odd numbers.

The next two terms are 35 + 13 = 48 and 48 + 15 = 63

You can use the same method to work out missing terms within a sequence.

What are the missing terms in this sequence?   1, ..., 4, 8, ..., 32

Write out the sequence with the differences.

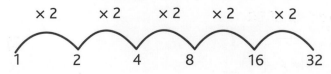

Number puzzles

Number puzzles are often written in words. You can solve these by writing statements first – use the same methods that you used to write statements to solve problems in Section A.

Two numbers have a sum of 11 and a product of 24. What are the numbers?

Use the information you are given to write some statements.
Use letters or symbols to represent the unknown numbers, for example, a and b.

$a + b = 11$ $a \times b = 24$

Write down the factor pairs of 24: $1 \times 24, 2 \times 12, 3 \times 8, 4 \times 6$

Which of these factor pairs satisfies the first statement? $a + b = 11$
$3 + 8 = 11$

Therefore $a = 3$ and $b = 8$ (or vice versa).
So, 3 and 8 are the numbers.

> **Key words:** sequence, rule, pattern, function, term

Train

1   Write down the next three terms in these 'counting on' and 'counting back' sequences.   (5)
    (a) 1, 6, 11, 16, 21, ..., ..., ...   *26, 31, 36*   (b) 47, 40, 33, 26, 19, ..., ..., ...   *12, 5, -2*   (c) 4.0, 4.2, 4.4, 4.6, 4.8, ..., ..., ...   *5, 5.2,*

Try

2   Write down the next two numbers in each sequence.   (4)
    (a) 11   19   27   35   43   ...   ...   *51, 59*   (c) 15   12   9   6   3   ...   ...
    (b) 100   93   86   79   72   ...   ...

3   Write down the missing terms in these sequences.   (2)
    (a) ..., ..., 15, 20, 25   (b) ..., 10, 12, 15, 19, ..., 30

Test

4   Write down the next two terms in these sequences.   (5)
    (a) 2   3   6   11   18   ...   ...   (c) 2   5   11   23   ...   ...
    (b) 320   160   80   40   20   ...   ...

5   Write down the missing terms in these sequences.   (4)
    (a) 0.5, ..., 1, 1.25, ..., 1.75   (b) 16, ..., 36, ..., 64, 81

6   Two numbers have a product of 24 and a sum of 10
    What are the two numbers?   (2)

7   Two numbers have a difference of 11 and a product of 60
    What are the two numbers?   (2)

E: Equations

About this topic: In the previous section, you revised writing formulae using symbols. In this section, you will work more with formulae and extend your revision to other types of equations.

In the last section, you used flow charts to represent word formulae with letters to represent the unknowns. You can take this one step further and write an equation.

An **equation** consists of two expressions that are equal to each other that is, they **balance**. One expression can simply be a number, such as a result.

Word formula: Ynes thought of a number, subtracted 4 and then multiplied by 3, her result was 27

As an equation: $3(c - 4) = 27$ where c is the number Ynes first thought of.

Solving equations

You can solve an equation to find the value of the unknown number(s). In order to keep the balance, you have to do the same thing to both sides of the equation.

Solve: $a + 3 = 7$

$a = 7 - 3 = 4$ Subtract 3 from both sides to find the value of a

Solve: $b - 4 = 12$

$b = 12 + 4 = 16$ Add 4 to both sides to find the value of b

Solve: $4c = 12$

$c = 12 \div 4 = 3$ Divide both sides by 4 to find the value of c

It takes more than one step to solve some equations and find the value of the unknown.

Solve $2d + 3 = 7$

$2d = 7 - 3 = 4$ Subtract 3 from both sides to find the value of $2d$

$d = 4 \div 2 = 2$ Divide both sides by 2 to find the value of d

Solve $2(e + 3) = 16$

$2e + 6 = 16$ Multiply out the bracket first.

$2e = 10$ Subtract 6 from both sides.

$e = 5$ Divide both sides by 2

> **Key words:** equation, expression, balance

Train

1 For each part, write the word formula as an equation and then solve it to find the unknown number.

(a) Sam is 11 years old. He is 2 years younger than Maya. How old is Maya? (1)

(b) A regular hexagon has a perimeter of 42 cm. What is the length of one side of the hexagon? (1)

2 What numbers are represented by the symbols in these equations?

(a) $7 + \blacksquare = 23$ (1)

(b) $\blacklozenge - 8 = 7$ (1)

(c) $9 \times * = 36$ (1)

3 What are the values of the letters in these equations?

(a) $a + 5 = 12$ (1)

(b) $b - 6 = 6$ (1)

(c) $4 \times c = 20$ (1)

Try

4 What numbers are represented by the symbols in these equations?

(a) $5 \times \blacksquare \times 7 = 140$ (1)

(b) $37 + 44 = * \times 9$ (2)

5 What are the values of the letters in these equations?

(a) $3d - 4 = 2$ (2)

(b) $3(e - 4) = 9$ (3)

Test

6 What are the values of the letters in these equations?

(a) $v + 6 = 20$ (1)

(b) $2w - 8 = 40$ (2)

(c) $5(x + 6) + 5 = 95$ (3)

7 Chris thought of a number, added 5 and multiplied the result by 8. His final result was 48. What was the number her first thought of? (1)

8 Three friends, Louis, Archie and India, each thought of number.

(a) Write an expression for the sum of their numbers. (1)

(b) If Louis' number was 2, Archie's number was 5 and India's number was 7, what was the sum of their numbers? (1)

(c) If Archie's number was 6, India's was 12 and the sum of their three numbers was 35, what was Louis' number? (2)

Test 5

1   Izzy has made a machine which adds 4 to every input number.

input → [+ 4] → output

   (a) Izzy puts 9 into the machine. What is the output? (1)

   (b) If the output is 4, what number did Izzy put into the machine? (1)

2   Abdul has made a machine. His machine multiplies by 3 and then subtracts 5

input → [× 3] → [− 5] → output

   (a) Abdul puts 3 into the machine. What is the output? (1)

   (b) If the output is ⁻2, what number did Abdul put into the machine? (1)

3   Write (i)   the function involved   and   (ii)   the next two terms (4)

   in each sequence.

   (a) 3, 6, 9, 12, 15, ..., ...   (c) 37, 33, 29, 25, 21, ..., ...

   (b) 1, 2, 4, 8, 16, ..., ...   (d) 22, 18, 14, 10, 6, ..., ...

4   Write the next two numbers in each sequence. (2)

   (a) 1, 2, 4, 7, 11, 16, ..., ...   (b) 1, 3, 4, 7, 11, 18, ..., ...

5   Two numbers have a sum of 32 and a product of 60. What are the numbers? (2)

6   I think of a number less than 20. It has no remainder when divided by 3 but the remainder
   is 3 when it is divided by 4. What is the number? (1)

7   Anne thought of a number and added 9. The result was 23. What number did Anne think of? (1)

8   Write these word sentences as formulae. Use brackets where appropriate. (2)

   (a) Divide 6 by d (b) Take f away from g and multiply the result by h.

9   What is the value of x in this equation? $6(x − 3) + 8 = 44$ (2)

10 (a) Draw the next two patterns in this sequence. (2)

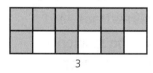

   1   2   3

   (b) Copy and complete this table for the sequence. (3)

Pattern number	Pink squares	White squares	Total squares
1	3	1	4
2	6		
3			
4			
5			

   (c) Write a formula for the total number of squares in the nth term of this sequence. (1)

   (d) Write a formula for the total number of pink squares in the nth term of this sequence. (1)

**Record your score and time here and
at the start of the book.**

Score [   ] / 25   Time [   ] : [   ]

⑥ Handling data

Introduction

In everyday life, you are surrounded by lots of information or data, presented in many different forms, including lists, charts, graphs and spreadsheets. In this chapter, you will look at displaying and interpreting different types of data.

In order to display and understand data properly, you need to understand what type of data you are dealing with. **Discrete data** is usually associated with counting, using integers.

The number of cars passing the school gates is an example of discrete data. Points in between the 'whole' numbers of cars have no meaning. You can't count 2.5 cars!

Discrete data is often displayed using **Carroll diagrams**, **Venn diagrams**, **pictograms**, **bar charts**, **pie charts** and **frequency diagrams**.

Continuous data is usually associated with measurement.

Heights and temperatures are both examples of continuous data. They change gradually over time.

Continuous data is most accurately represented by a **line graph**.

In this chapter, you will revise the various ways of representing data.

When you **collect data**, you can record it in a **list**, a **tally table** or a **frequency table**. You may have used these methods when recording the results of experiments in science or weather data in geography.

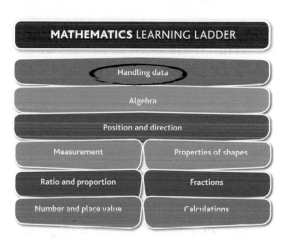

Congratulations! Once you have completed the chapter, you have reached to the top of the mathematics Learning Ladder. You have revised all the maths skills you need for your 11+ tests and exams.

Revisit any areas that you have found challenging. Re-read the notes and work through the examples and the questions until you can complete them confidently.

Advice for parents

Data handling is a very visual, and therefore accessible, branch of mathematics. Ensure a good understanding of the different methods used to record, display and analyse data.

Before starting work on this chapter, it might be useful to revisit related information from earlier chapters, for example:
- angle sum at the centre of a circle on page 90 (drawing of pie charts)
- calculating percentages on page 62 (working out the segment sizes in a pie chart)
- fractions in their simplest form on page 42 (pie charts)
- fractions, percentages and ratios on page 46 (probability).

Revisit any relevant definitions that cannot be recalled readily and confidently, for example, range, mode, median and mean.

A sharp pencil is needed to draw bar charts, frequency diagrams, pie charts and line graphs. A protractor, a ruler and compasses are also essential items.

A: Carroll diagrams and Venn diagrams

About this topic: You will begin your data handling revision by looking at two simple ways of representing discrete data.

Carroll diagrams

The Carroll diagram was named after Lewis Carroll, the author of *Alice in Wonderland*! It is used to sort and group data according to certain criteria.

The table and the Carroll diagram show the same information about the 'vehicles' owned by a group of pupils.

Vehicles owned	Number of pupils
A bicycle only	✔ ✔ ✔ ✔
A bicycle and a scooter	✔ ✔ ✔ ✔
A scooter only	✔ ✔ ✔
Neither a bicycle nor a scooter	✔

■ Carroll diagrams

	Bicycle	No bicycle
No scooter	4	1
Scooter	4	3

The Carroll diagram shows clearly that a total of 8 pupils own a bicycle and 4 pupils own both a bicycle and a scooter.

Venn diagrams

A Venn diagram uses overlapping circles or loops to show how groups are related. The area where the circles overlap contains numbers that relate to the properties of both groups.

This Venn diagram displays the same information as the table and the Carroll diagram above.

Notice how the regions of the Venn diagram and Carroll correspond. The 1 outside the circles but inside the rectangle represents the pupil who does not own a scooter or a bicycle.

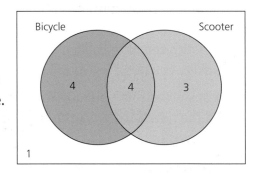

Venn diagrams can also be used to find the **highest common factor** of two or more numbers. Each circle in the Venn diagram shows all the numbers in the prime factorisation of that number. The region where the circles overlap show the shared prime factors. The highest common factor is the product of all these common prime factors.

Use a Venn diagram to find the highest common factor of 12 and 18

$12 = 2 \times 2 \times 3$

$18 = 2 \times 3 \times 3$

2 and 3 are common in the prime factorisation of 12 and 18

So, the highest common factor is $2 \times 3 = 6$

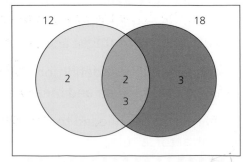

Key words: discrete data, Carroll diagram, Venn diagram, highest common factor, common factor

Train

1 This Venn diagram shows the numbers of boys in a
   class who play cricket and tennis in the
   summer term.   (7)

   (a) How many boys play tennis but not cricket? 4

   (b) How many boys play neither tennis nor cricket? 3

   (c) How many boys are in the class? 12

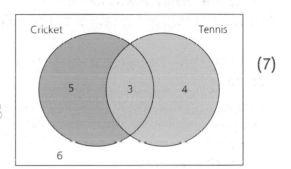

Try

2 Look at this set of numbers.   (8)

1	2	3	4	5	6	7	8	9	10
11	12	13	14	15	16	17	18	19	20

   (a) Write each number in the correct region
   of this Carroll diagram.

   (b) Draw a Venn diagram showing those numbers that
   are odd and those that are divisible by 3
   Write each number in the correct region.

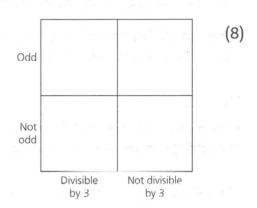

Test

3 The table shows the results of a survey about siblings for a group of year 6 pupils.   (8)

Siblings	Number of pupils
Brothers only	✔ ✔ ✔ ✔ ✔
Sisters only	✔ ✔ ✔ ✔ ✔ ✔ ✔ ✔
Sisters and brothers	✔ ✔ ✔ ✔ ✔ ✔ ✔ ✔ ✔ ✔
No brothers or sisters	✔ ✔ ✔

   (a) Write the results in the correct region of this
   Carroll diagram.

   (b) Represent this data in a Venn diagram with
   two sets: Brothers and Sisters.

4 **(a)** Write 16 as the product of its prime factors.   (2)

   (b) Write 24 as the product of its prime factors.   (2)

   (c) Draw a Venn diagram to represent the information in parts (a) and (b).   (2)

   (d) What are the shared prime factors of 16 and 24?   (1)

   (e) What is the highest common factor of 16 and 24?   (1)

B: Pictograms

About this topic: In this section you will revise an alternative way of displaying discrete data, pictograms.

A **pictogram** uses simple pictures or symbols to represent one or more pieces of discrete data.

Reading pictograms

You always need to check what one symbol represents when you are reading a pictogram.

■ The numbers of boys and girls standing in a line

Boys	☺ ☺ ☺ ☺
Girls	☺ ☺ ☺ ☺ ☺ ☺ ☺ ☺

Key: ☺ represents 1 child

The key shows that in this pictogram one symbol represents one child. There are 4 boys and 8 girls standing in the line.

A symbol can represent any number or type of items, for example, 2 children, 10 chickens or 1 million pounds. If the key indicates that ☺ represents 2 people, then ☾ represents 1 person.

> When reading a key, always check carefully what the symbol represents.

Drawing pictograms

Follow these steps to draw a pictogram.

● Write a title at the top of your pictogram.
● Choose an appropriate symbol or picture.
● Include a key. Be very clear about how many items your symbol represents.

> **Key words:** pictogram, discrete data, key

Train

1 The pictogram below shows the number of marks achieved by a group of pupils in a spelling test.

■ Marks in a spelling test

Oriel	✓ ✓ ✓ ✓ ✓ ✓ ✓	7
Sam	✓ ✓ ✓	3
Chris	✓ ✓ ✓ ✓ ✓ ✓	6
Maya	✓ ✓	2
Edward	✓ ✓ ✓ ✓	4

Key: ✓ represents 1 mark

(a) Who scored the highest mark? (1)

(b) Who scored the lowest mark? (1)

(c) What is the difference between the highest and lowest scores? (1)

(d) A total of eight marks was available. How many pupils scored more than half marks? (1)

Try

2 The pictogram below shows the money raised for charity by four friends.

■ Money raised for charity

Toby	£	£					*10*
Max	£	£	£				*15*
Camilla	£	£	£	£			*20*
Dorothy	£	£	£	£	£	£	*30*
Esther	£	£	£	£	£		*25*

Key: £ represents £5

(a) Who raised the most money? (1)

(b) How much more money did Dorothy raise than Toby? (1)

(c) How much money was raised altogether? (1)

(d) Which two people raised £55 between them? (1)

Test

3 The chart shows the numbers of bottles of water sold from the vending machine during one week.

■ Bottles of water sold in one week

Monday	● ● ● ●	*16*
Tuesday	● ● ● ● ⸰	*18*
Wednesday	● ● ●	*12*
Thursday	● ● ● ● ● ●	*24*
Friday		*No*

Key: ● represents 4 bottles

(a) How many bottles of water were sold on Thursday? (1)

(b) How many bottles of water were sold on Tuesday? (1)

(c) 20 bottles of water were sold on Friday. Copy the chart and fill in the results for Friday. (2)

(d) What was the total number of bottles sold that week? (2)

C: Bar charts

Bar charts are a very useful way to display discrete data. In this section, you will revise how to read information from bar charts and practise drawing them.

A **bar chart** displays data using horizontal or vertical rectangular bars.

Reading bar charts

The labels on the axes will tell you what information is being displayed. The height or length of the bar might, for example, record the number of objects in a particular group.

This bar chart shows the numbers of girls and boys standing in a line.

The vertical axis shows the number of children.

Each division represents 1 child.

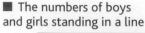
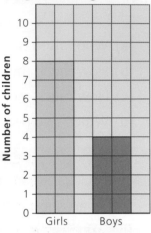

■ The numbers of boys and girls standing in a line

Look carefully at the scale when you read a bar chart.

This bar chart shows the heights of runner bean plants after two weeks.

Each division represents 5 cm

You may have drawn a bar chart like this in a science lesson.

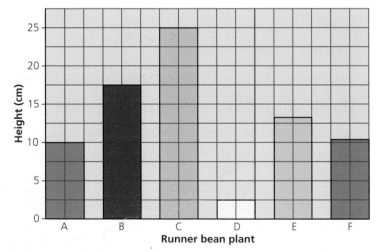

■ The heights of runner bean plants after two weeks

Drawing a bar chart

Follow these steps when you draw a bar chart.

● Write a title at the top of your bar chart.
● Choose an appropriate scale. It will usually start at 0
● Label both axes.
● Draw all your bars the same width.

Key words: bar chart, scale

Train

1 This bar chart shows the numbers of different types of pet owned by pupils in Year 6

(a) How many dogs are owned by pupils in Year 6?

(b) How many cats are owned by pupils in Year 6?

(c) How many more cats are owned than rabbits?

(d) How many pets do Year 6 pupils own in total?

(6)

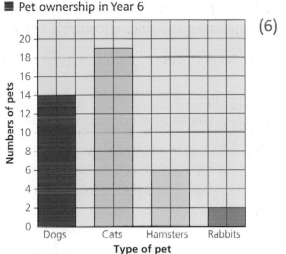

■ Pet ownership in Year 6

Try

2 Archie's cat has had six kittens. The bar chart shows the masses of five of the kittens.

(a) What is the mass of kitten C?

(b) What is the mass of kitten A, in kilograms?

(c) The total mass of the six kittens is 900 g. Calculate the mass of kitten F, in grams.

(d) Copy the bar chart and draw the bar to show the mass of kitten F.

(8)

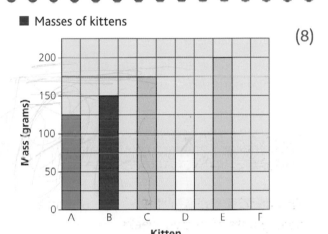

■ Masses of kittens

Test

3 The bar chart shows the numbers of ice creams sold over a one-year period.

(6)

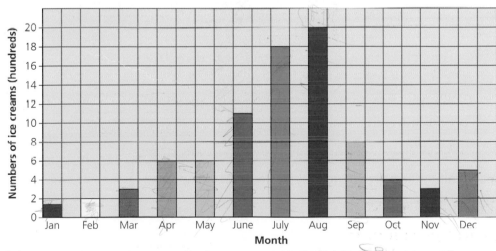

■ Sales of ice creams

(a) In which month were the least ice creams sold? *Feb*

(b) In which month were the most ice creams sold? *Aug*

(c) (i) In which two months were the same number of ice creams sold? *Apr and May*

(ii) How many ice creams were sold altogether during these two months? *6*

(d) The sales of ice creams increase leading up to August. Suggest a reason for this.
ice creams are brough more often in the hot hot(t) summer months

D: Frequency diagrams

About this topic: Frequency means how often things happen. In this section you will revise how to produce a frequency table from a tally chart and how to display the results from a frequency table as a frequency diagram.

Tally charts and frequency tables

In a tally, the **tally marks** are recorded in blocks of 5 like this:

| ‖ ‖‖ ‖‖‖ ‖‖‖‖ 卌 卌|

1 2 3 4 5 6

You can turn a **tally chart** into a **frequency table** by adding a frequency column. The frequency for each score is the sum of the tallies.

Score	Tally	Frequency
1	卌 卌 卌	15
2	卌 卌 卌 ‖‖‖	19
3	卌 卌 卌	16
4	卌 卌 卌 卌	20
5	卌 卌 ‖‖‖	13
6	卌 卌 卌 ‖	17
	Total	**100**

■ Scores when rolling a die 100 times

Frequency diagrams

A frequency diagram is a type of bar chart. Frequency is always one of the two scales on a frequency diagram.

It shows clearly the result that occurs the most (and least) often.

This frequency diagram shows the information from the frequency table in the first example.

■ Scores when rolling a die

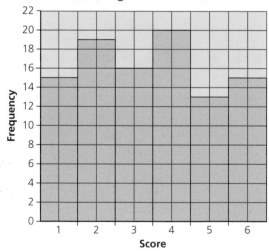

Key words: tally chart, tally marks, frequency table, frequency diagram

Train

1   The school football team played 15 matches this term. The number of goals they   (7)
    scored in each match are shown below.

    1 0 2 3 2 2 4 5
    0 1 2 3 0 1 2

    (a) Copy and complete the table for
        the number of goals scored.

    (b) Draw a frequency diagram to show
        the information in the table.

Goals scored in a match	Tally	Frequency
0	III	3
1		
2		
3		
4		
5		
	Total	

Try

2   This frequency diagram shows the shoe sizes of pupils in a Year 4 class.

■ Shoe sizes in Year 4

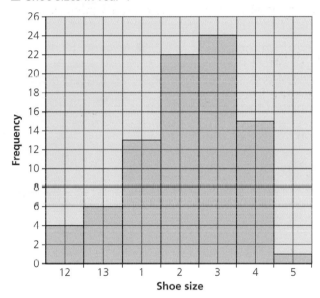

    (a) Which is the most common shoe size?   (1)
    (b) How many pupils have shoes size 12 or 13?   (2)
    (c) How many pupils have shoe size 1 or larger?   (1)
    (d) How many pupils are there in the class?   (1)

Test

3   The list below shows the test scores (out of 10) of a group of pupils.

    5   3   8   6   3   10   9   7   7   5
    10 4   8   7   6   7   8   4   2   9

    (a) Record these scores in a frequency table with a tally column.   (5)
    (b) Draw a frequency diagram to show the information.   (5)
    (c) How many test scores were recorded in total?   (1)
    (d) How many pupils achieved more than half marks?   (1)

E: Pie charts

About this topic: In this section you will revise another method of displaying data in a visual way: pie charts.

Pie charts show each category (part of the data) as a proportion of the whole set of data. They are useful for comparing quantities.

A pie chart is a circle that is divided into **sectors**. Each sector represents one category of the data. The sectors can be expressed as **fractions** or **percentages**.

This pie chart shows the data from 6C (page 138) about the numbers of boys and girls standing in a line.

8 of the 12 children are girls. As a fraction, the number of girls is $\frac{8}{12} = \frac{2}{3}$

4 of the 12 children are boys. As a fraction, the number of boys is $\frac{4}{12} = \frac{1}{3}$

$\frac{2}{3}$ of the pie is labelled 'girls' and $\frac{1}{3}$ is labelled 'boys'.

■ The numbers of boys and girls standing in a line

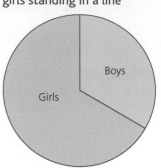

This pie chart shows the results of a survey into favourite foods.

The largest sector is pasta, so this is the most popular food.

The smallest sector is peas, so this is the least favourite food.

$\frac{1}{4}$ (or 25%) of the people liked chocolate best. (It takes up $\frac{1}{4}$ of the circle.)

If you were told that 36 people took part in the survey altogether, you can work out the number of people that chose each type of food.

$\frac{1}{4}$ of the people chose chocolate as their favourite food, so $\frac{1}{4} \times 36 = 9$ people.

■ Favourite foods

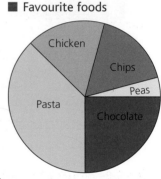

> **Key words:** pie chart, sectors, fractions, percentages

Train

1   This pie chart shows the results of a survey about car ownership.

(a) What fraction of households surveyed have only one car?

(b) What percentage of households surveyed have only one car?

(c) What percentage of households surveyed have more than one car?

(d) The survey was completed by 200 households. How many of the household have only one car?

■ Number of cars in a household   (5)

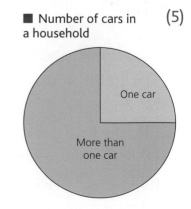

Try

2 In the autumn term, pupils in Class 6 choose between playing rugby, soccer, netball or hockey. This pie chart displays the sports chosen by Class 6 last year.

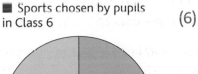
■ Sports chosen by pupils in Class 6

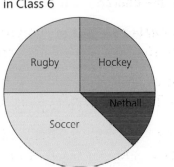

(a) What percentage of Class 6 chose rugby?

(b) What fraction of Class 6 chose netball?

(c) What fraction of Class 6 chose soccer?

(d) 12 pupils chose rugby. How many pupils were in Class 6?

Test

3 The pie chart shows information about the spiders and insects found on a nature walk.

■ Spiders and insects found on a nature walk

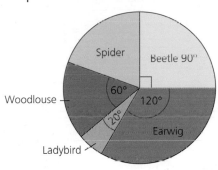

(a) What fraction of the pie chart represents beetles? (1)

(b) What fraction of the pie chart represents earwigs? (1)

(c) Calculate the size of the angle that represents spiders. (1)

(d) What fraction of the pie chart represents spiders? (1)

(e) 72 minibeasts were found in total.

   (i) How many beetles were found? (2)

   (ii) How many earwigs were found? (2)

   (iii) How many ladybirds were found? (2)

> The angle sum at the centre of a circle is 360°

F: Drawing pie charts

About this topic: In the last section, you revised how to read information from pie charts. In this section, you will practise drawing pie charts.

In the previous section, you wrote categories or sectors of a pie chart as fractions of the whole (pie):

$$\frac{\text{the number of items in a category}}{\text{the total number of items}}$$

To draw a pie chart, you need to know the size of each sector. This is the fraction of 360° that will represent each category in the pie chart.

The frequency table shows the main instruments played by 20 pupils. Two extra columns have been added: the first shows the fraction of people that play each instrument; the second shows how to use the fraction to calculate the angle of the sector that will represent this category.

Instrument	Tally	Frequency	Fraction	Angle in pie chart
Piano	JHT IIII	8	$\frac{8}{20} = \frac{2}{5}$	$\frac{2}{5} \times 360° = 144°$
Guitar	IIII	4	$\frac{4}{20} = \frac{1}{5}$	$\frac{1}{5} \times 360° = 72°$
Drums	II	2	$\frac{2}{20} = \frac{8}{10}$	$\frac{1}{10} \times 360° = 36°$
Violin	JHT I	6	$\frac{6}{20} = \frac{3}{10}$	$\frac{3}{20} \times 360° = 108°$
Total		**20**		**360°**

Once you have calculated the angles of the sectors to represent the number of people in each category, you can draw a pie chart.

Follow these steps to draw a pie chart.

- Use compasses to draw a circle.
- Draw a line from the centre of the circle to the circumference.
- Then use your protractor to measure and mark each angle. Draw a straight line through the mark, using a ruler and a sharp pencil.
- Colour and label each sector.
- Give your pie chart a title.

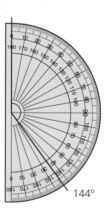

144°

This pie chart shows the data from the example above.

■ Instruments played

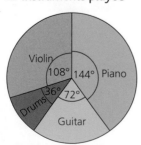

Key words: pie chart, 360° (angle sum at the centre of a circle)

Train

1 The table shows the results of a survey of pupils' favourite fruit. Copy and complete the table to show the angles that would represent each fruit in a pie chart. (10)

Fruit	Frequency	Fraction	Angle in pie chart
Banana	5		
Apple	9		
Satsuma	6		
Grapes	10		
Total	30		

Try

2 Draw a pie chart to show the information in the table in question 1 (5)

Test

3 24 people were asked to name their preferred holiday destination. The frequency table shows the results of the survey.

(a) Copy and complete the frequency table to show the angle that would represent each destination in a pie chart. (10)

Destination	Frequency	Fraction of 360°	Angle
United Kingdom	8		
France	4		
Italy	3		
Spain	4		
Greece	5		
Total	24		

(b) Draw a pie chart to show the information. (5)

(c) Which was the least popular destination? (1)

(d) Which was the favourite destination? (1)

G: Range, mode, median and mean

About this topic: In this section, you will move on to compare two or more sets of data with numerical values. To do this, you need to be able to calculate an average that represents the data and the range of the data.

All the examples in this section relate to this ordered data set.

■ Test scores (out of 10) achieved by 12 children

3	4	5	6	7	7	8	8	8	8	10	10

Read through each subsection in turn and work through the calculation. This should help you to remember the meaning of the each term and how to calculate it.

Range

The **range** of a set of data is the **difference** between the largest and smallest values.

Find the largest value (10) and the smallest value (3)

Range = 10 − 3 = 7

Mode

The **mode** (modal value) is the value that occurs **most often** in the data set.

There are more 8s than any other number in the data set, so the mode is 8

Median

The **median** of a set of data is the **middle value** of the set when it is arranged in numerical order.

If there is an even number of items in the data set, there will be two middle values. The median value is half way between these two values.

3   4   5   6   7   ⑦   ⑧   8   8   8   10   10

The data set has two middle values: 7 and 8

7.5 is half way between 7 and 8, so the median is 7.5

Mean

The **mean** of a set of values is found by adding all the values and then dividing by the number of values.

$$\text{Mean} = \frac{\text{the total of all values}}{\text{the number of values}}$$

Add up all the scores.   3 + 4 + 5 + 6 + 7 + 7 + 8 + 8 + 8 + 8 + 10 + 10 = 84

Count the number of scores. 12

Mean = 84 ÷ 12 = 7

> **Key words:** average, range, mode, median, mean

Train

1 Write each set of numbers in order, starting with the smallest. (12)
   (a) 1 5 2 4 3 6 5 6 4 4
   (b) 29 34 23 45 25 43 34 55
   Then calculate:
   (i) the range   (ii) the mode   (iii) the median   (iv) the mean.

Try

2 These spelling test scores are arranged in numerical order.

   3 3 4 4 5 5 6 6 6 7 7 7 7 8 8 10

   (a) What is the range? (1)
   (b) What is the mode? (1)
   (c) What is the median score? (1)
   (d) Calculate the mean. (2)

3 These are the overnight temperatures (°C) each day during one week in June.

   15 13 10 10 13 12 14 9

   What is the mean overnight temperature for this week? (2)

Test

4 This table shows the results of a class survey about how much pocket money children received each week.

Amount of pocket money	Number of children (frequency)
£0	1
£1	6
£2	7
£3	3
£4	2
£5	4
£6	1
£7	1

   (a) Draw a frequency diagram to show this data.
       (Drawing a frequency diagram was covered in part D.) (5)
   (b) How many children took part in the survey? (1)
   (c) What is the range of the data? (1)
   (d) What is the modal value? (1)

   This list shows the amounts of pocket money in numerical order:

   0 1 1 1 1 1 1 2 2 2 2 2 2 2 3 3 3 4 4 5 5 5 5 6 7

   (e) What is the median amount of pocket money? (2)
   (f) What is the mean amount of pocket money? (2)

Grouped data

In this topic: The data sets in the previous sections have all been quite small. In this section you will revise how best to work with larger amounts of data.

If there is a lot of data, it is usually more practical to organise the data in suitable **groups** or **class intervals**.

If, for example, you want to represent results of an examination marked out of 100, it would take a long time to draw a frequency diagram with 100 bars.

It is more practical to group the data into suitable intervals, such as:

0–9   10–19   20–29   30–39   and so on.

The data below shows the numbers of goals scored by the A team in each netball match they played last season.

```
 1    3    6    7    9
 1    7    4   11    8
 3   12   11    8   10
10   11   10   13    4
 5   15    9   10   12
```

Draw a frequency table for the data. Use suitable intervals.

You can use any equal grouping that gives a reasonable number of categories.
One possibility is:

Number of goals scored	Tally	Frequency
1–3	IIII	4
4–6	Ⅲ̶Ⅰ	5
7–9	Ⅲ̶Ⅰ I	6
10–12	Ⅲ̶Ⅰ III	8
13–15	II	2
	Total	**25**

You can now draw a Frequency diagram

■ Number of goals scored

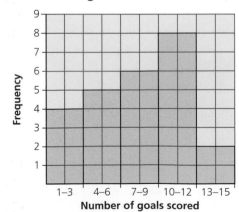

Number of goals scored

Key words: grouped data

Train

1 The children in Mrs Willett's class have made this list of their waist measurements in centimetres. This is shown below:

61 72 68 63 69 58 67 62 69 57

70 61 70 78 67 65 63 62 64 75

(a) What is the range of the data? (2)

(b) What is the median? (3)

(c) Copy and complete the grouped frequency table for this data. (4)

Waist measurement (cm)	Tally marks	Frequency
55–57		
58–60		
61–63		
64–66		
67–69		
70–72		
73–75		
76–78		
	Total	

Always write the data in order first.

Try

2 (a) Draw a frequency diagram to show the data in question 1 (4)

(b) Which is the modal group? (1)

Test

3 Look at the data about the number of goals Team A scored in each netball match in the example on the previous page.

(a) Draw a frequency diagram to show the information. (5)

(b) What was the modal number of goals scored? (1)

(c) What is the mean number of goals scored? (1)

(d) In how many matches did Team A score fewer than 10 goals? (1)

I: Line graphs

About this topic: The charts and tables you have revised so far are usually used to display discrete data. In this section, you will look at how to display and read line graphs that are used to represent **continuous data**.

A **line graph** can show how something changes over time, for example, temperature, height, depth or distance.

As with other types of graphs, it is very important that you understand the **scale** before you read information from a line graph.

Travel graphs

Line graphs can be used to show journeys. These are sometimes called **travel graphs** or **distance–time graphs**.

Horizontal lines on these graphs mean that the person or vehicle is not moving at that time.

Robert cycled from his house to his friend's house and back. The graph shows his distance from home during this time.

Robert's journey to his friend's house was 9 km.

4 squares on the horizontal axis represent 1 hour (60 minutes), so 1 square on the horizontal axis represents 60 ÷ 4 = 15 minutes.

He stopped for 15 minutes on the way to his friend's house (between 10:00 and 10:15).

Robert spent 30 minutes at his friend's house before he cycled home.

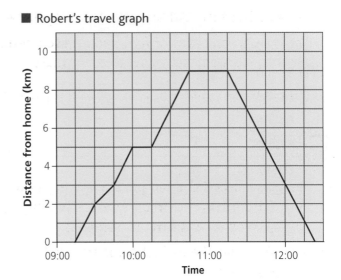

■ Robert's travel graph

Other line graphs

Line graphs are also used for other continuous measures, such as growth, temperature and the depth of liquid in a container over time.

> **Make sure you understand the scale.**

This line graph shows the depth of water in a water tank.

Water was taken from the tank after 1 hour and again $2\frac{1}{2}$ hours after the tank was filled.

Both times water was taken, the water level fell 0.5 feet.

After the second time water was taken from the tank, the water level continues to fall slowly. Perhaps the tap was left dripping!

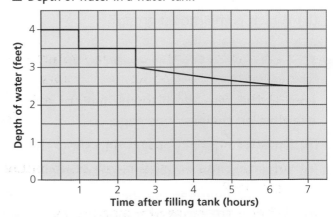

■ Depth of water in a water tank

Key words: line graph, continuous data, scale

Train

1 This line graph shows the amount of fuel in Stanley's car during a car journey

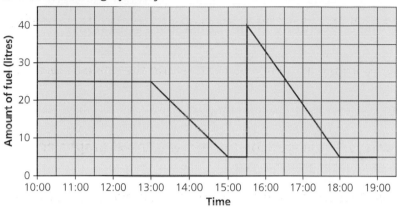

■ Fuel used during a journey

(a) At what time did Stanley start his journey? (1)

(b) How many litres of fuel had he used before he stopped to refuel? (1)

(c) At what time did he fill up with fuel? (1)

(d) How many litres of fuel did Stanley use for his whole journey? (2)

Try

2 Bertie recorded the temperature every hour for 12 hours on the first day of his holiday. The graph shows his measurements. (6)

(a) What was the lowest temperature he measured?

(b) At what time was the temperature 28 °C?

(c) What was the range of temperatures?

(d) What was the mean temperature?

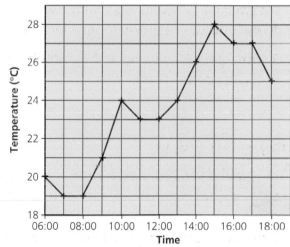

■ Hourly temperatures

Test

3 Mrs Constable recorded the temperature in her greenhouse every hour for 15 hours during one day in February. The line graph shows her results. (4)

(a) What was the temperature at noon?

(b) At what time was the lowest temperature recorded?

(c) What is the range of temperatures during the 15-hour period?

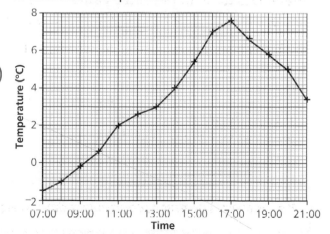

■ Greenhouse temperatures

J: Drawing line graphs

About this topic: Now that you are confident with reading line graphs, you can practise drawing them.

There are a number of important factors to consider when you draw (or read) a line graph.

Look again at the line graph showing the temperature in Mrs Constable's greenhouse.

When you draw a line graph, you must include a **title**. When you read a line graph, the title gives important information about what the graph shows.

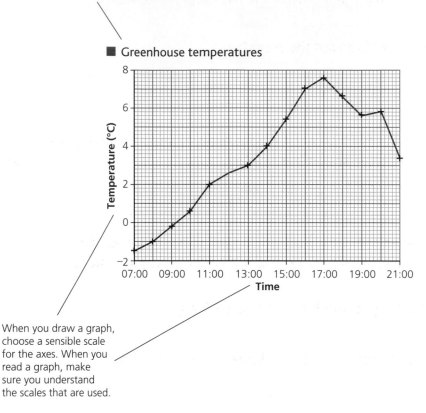

■ Greenhouse temperatures

When you draw a graph, choose a sensible scale for the axes. When you read a graph, make sure you understand the scales that are used.

Follow these steps when you draw a line graph:

● Use graph paper or square paper.
● Use a sharpened pencil and ruler.
● Choose a sensible scale for your axes (vertical and horizontal).
● Draw and label your scales.
● Write a title.
● Plot all your points with a small (+).
● Use a ruler to help you draw straight lines between the points.

Follow these steps as you work through the Train, Try, Test questions.

Key words: title, horizontal axis, vertical axis, scale

Train

1 The mass of a kitten was recorded over an 8-week period. (8)

Week	1	2	3	4	5	6	7	8
Mass (grams)	90	180	200	240	260	290	300	330

(a) Draw a set of axes on graph paper. Label the horizontal axis 'Week' and mark it in intervals of 1 week. Label the vertical grid 'Mass (grams)' and mark it in intervals of 50 g.

(b) Plot the recorded masses on the grid. Mark each point with a (+).

(c) Join up the points.

(d) In which week did the kitten's mass increase the most?

Try

2 Ella recorded the distance she had travelled towards school every 4 minutes during one morning's car journey to school. The table shows her results. (8)

Time (minutes)	0	4	8	12	16	20	24	28	32	36
Distance (km)	0	1	2	4	5	6	6	7	9	11

(a) Draw a set of axes on graph paper. Label the horizontal axis 'Time (minutes)' and mark it in intervals of 4 minutes. Label the vertical axis 'Distance (km) and mark it in intervals of 1 km.

(b) Plot the recorded distances on the grid. Mark each point with a (+).

(c) Join up the points.

(d) What do you think happened between 20 and 24 minutes after her journey had started?

Test

3 The temperature of a pan of water was recorded every minute as it was heated in a pan. (8)

Time (minutes)	0	1	2	3	4	5	6	7	8	9	10
Temperature (°C)	15	23	31	43	54	60	68	75	80	85	90

(a) Draw a set of axes on graph paper. Label the horizontal axis 'Time (minutes)' and mark it in intervals of 1 minute. Label the vertical axis 'Temperature (°C)' and mark it in intervals of 10 °C.

(b) Plot the recorded temperatures on the grid. Mark each point with a (+).

(c) This time, join the points with a smooth curve.

(d) What was the temperature of the water after 6.5 minutes?

K: Conversion graphs

About this topic: In this section, you will revise another type of line graph.

A **conversion graph** is a line graph that shows a **fixed relationship** between two **variables**, for example, between two currencies such as UK pounds and Euros.

You need two reference points to draw a conversion line graph.

This conversion graph shows the relationship between kilograms (kg) and pounds (lb).

1 kg is very nearly the same as 2.2 lb so you can use (0, 0) and (1, 2.2) to draw this graph.

■ Conversion of kilograms (kg) and pounds (lb)

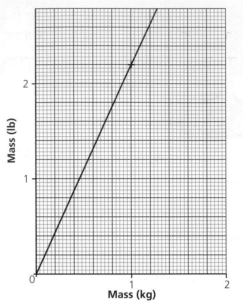

A shopkeeper bought a 25-metre roll of gold fabric for £18.75

He decides to sell the fabric (by the metre) for double the amount that it cost him, so £37.50 for the roll.

He draws a graph that he can use to work out what to charge a customer for any length of the fabric. The reference points are (0, 0) and (25, 37.50).

If a customer wants to buy 10 metres of fabric, the charge will be £15

If a customer has £30, they can buy 20 metres of the fabric.

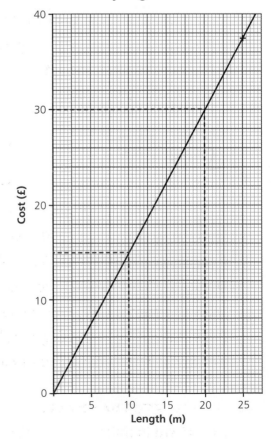

■ Cost of fabric by length

Key words: conversion graph, fixed relationship

Train

1   Use the shopkeeper's graph in the last example to answer these questions.
    (a) How much should the shopkeeper charge for 8 metres of gold fabric?   (2)
    (b) How much should the shopkeeper charge for 2.5 metres of gold fabric?   (2)
    (c) How much gold fabric could a customer buy for £15?   (2)
    (d) How much gold fabric could a customer buy for £5?   (2)

Try

2   This is a conversion graph between US dollars and UK pounds. ($1.60 = £1)

    ■ Conversion between pounds (£) and dollars ($)

    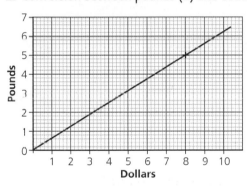

    (a) What does one small square represent on the vertical axis?   (1)
    (b) What does one small square represent on the horizontal axis?   (1)
    (c) What is £5 worth in dollars?   (1)
    (d) What is $9 worth in pounds?   (1)

Test

3   The conversion of miles to kilometres is 5 miles = 8 kilometres.
    (a) How many kilometres is 0 miles?   (1)
    (b) How many kilometres is 10 miles?   (1)
    (c) How many kilometres is 50 miles?   (1)
    (d) Draw a conversion graph for miles and kilometres on graph paper.

        Draw the horizontal axis using a scale of 1 cm to 10 km. Label it 'Distance in kilometres'.

        Draw the vertical axis using a scale of 1 cm to 10 miles. Label it 'Distance in miles'.

        Plot and join the points you calculated in parts (a)–(c). Make sure you plot them the right
        way round!   (5)

    (e) Use your graph to work out: (8)

        (i)   40 km in miles   (iii)   35 miles in kilometres

        (ii)   25 km in miles   (iv)   45 miles in kilometres.

L: Probability

About this topic: You have reached the final revision topic, **probability**. In this section, you will revise the likelihood of an event or events happening and a simple probability scale.

The **likelihood** that an **event** will happen can be:

certain	The event will always happen.	A pound coin is certain to sink if you drop it into water.
likely	The event will probably happen.	
even chance	The event has 'fifty-fifty' chance of happening.	A fair coin has an even chance of landing on heads.
unlikely	The event will probably not happen.	It is unlikely your teacher has blue hair.
impossible	The event will never happen.	It is impossible that the sun will rise in the west.

Outcomes

Every event has a number of possible **outcomes**. To work out the probability for an outcome, you need to know how many possible, equally-likely outcomes there are. It is helpful to list them.

When you toss a fair coin, there are two possible equally-likely outcomes: Heads and Tails.

When you roll a fair die, there are six possible equally-likely outcomes: 1, 2, 3, 4, 5, 6

When you spin this fair spinner, there are four equally-likely possible outcomes: 1, 2, 3, 4

The probability scale

The numbered probability scale goes from 0 (impossible) to 1 (certain).

```
0                                                               1
|_____|_____|
No chance            Even chance              Certain
```

Probability as a fraction

The probability of an event can be expressed as a fraction. The number of ways of getting the desired outcome is the **numerator** of the fraction. The number of possible outcomes is the **denominator** of the fraction.

The probability that a fair coin lands on Heads is: $\dfrac{1 \text{ Head}}{2 \text{ possible outcomes}} \rightarrow \dfrac{1}{2}$

The probability of scoring 3 with this fair spinner is: $\dfrac{1 \text{ way of scoring } 3}{4 \text{ possible outcomes}} \rightarrow \dfrac{1}{4}$

The probability of scoring 6 with a fair die is: $\dfrac{1 \text{ way of scoring } 6}{6 \text{ possible outcomes}} \rightarrow \dfrac{1}{6}$

Probability as a percentage

You can write a probability as a **percentage** by converting from a fraction.

As a percentage, the probability that a fair coin lands on Tails is: $\dfrac{1}{2} \times 100\% = 100 \div 2 = 50\%$

Probability as a ratio

You can also express probability as a ratio.

As a fraction, the probability of rolling a 4 with a fair die is: $\dfrac{1}{6}$
and as a ratio, it is 1:6, because rolling a 4 is one of 6 possible outcomes.

Outcomes of more than one event

You can use the same method to work out the probability of more than one event happening at the same time. Use a table to show all the possible outcomes.

This table shows the possible outcomes when two coins are thrown.

	Second coin – Heads (H)	Second coin – Tails (T)
First coin – Heads (H)	HH	HT
First coin – Tails (T)	TH	TT

The probability of throwing two Heads is: $\dfrac{1 \text{ way of two Heads}}{4 \text{ possible outcomes}} \rightarrow \dfrac{1}{2}$

The probability of throwing one Head and one Tail is: $\dfrac{2 \text{ ways for 1 Head and 1 Tail (HT, TH)}}{4 \text{ possible outcomes}} \rightarrow \dfrac{2}{4} = \dfrac{1}{2}$

Key words: probability, likelihood, event, outcome, probability scale, fraction

Train

1 (a) List all the possible outcomes when an ordinary die is rolled. (2)

   (b) What fraction of these outcomes: (i) are even numbers   (ii) is a 2? (2)

   (c) If you rolled the die 100 times, how many times might you roll an odd number? (1)

Try

2 Penelope has written the letters of her name on cards. (5)

| P | E | N | E | L | O | P | E |

She shuffles the cards and takes one card without looking. What is the probability that the letter on the card is: (a) a vowel   (b) P   (c) E   (d) a letter with reflection symmetry?

Test

3 Oriel has an ordinary coin and a pentagonal spinner. (10)

 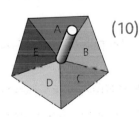

   (a) How many possible outcomes are there when she tosses the coin?

   (b) How many possible outcomes are there when she spins the spinner?

  Oriel tosses the coin and spins the spinner at the same time.

   (c) Draw a table to show all the possible outcomes.

   (d) How many possible outcomes are there?

   (e) What is the probability of getting a Head and the letter D?

   (f) What is the probability of getting Tails?

Test 6

1 There are 15 people standing outside the school gates. Each person has either a mobile phone, a coat or both. Ten people have a mobile phone and eight people have a coat. (2)

   (a) Copy and complete the Venn diagram to show this information.

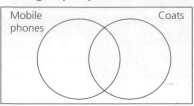

   (b) How many people had both a coat and a mobile phone?

2 The pictogram shows the number of boys and girls in each class in the junior school. (5)

   (a) What does the symbol ♦ mean?

   (b) How many boys are there in Year 6?

   (c) Which is the largest year group?

   (d) How many children are in the junior school?

   (e) How many more girls are there than boys?

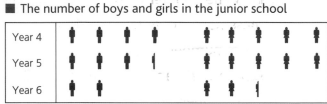

■ The number of boys and girls in the junior school

Key: ♦ represents 2 boys and ♦ represents 2 girls

3 The table shows the numbers of books read by six children over the summer holidays. (6)

Child	1	2	3	4	5	6
Number of books	2	3	4	6	6	9

   (a) What is the range?

   (b) What is the median number of books?

   (c) What is the mode?

   (d) What is the mean number of books read?

   (e) Draw a bar chart to display these results.

4 The frequency table records the results of the ages of the children in a cricket club.

Age	7	8	9	10	11	12	13	14
Frequency	2	4	5	8	4	3	6	3

   (a) How many children are in the cricket club? (1)

   (b) What is the range of their ages? (1)

   (c) How many children could play in a match for children aged 11 and over? (1)

   (d) What is the modal age? (1)

   (e) Draw a frequency diagram to show this information. (2)

5 Louis has two ordinary dice. He rolls both dice at the same time.

   (a) Draw a table to show all the possible outcomes. (2)

   (b) (i) List all the different ways to score a total of 4 (1)

      (ii) What is the probability of scoring 4? (1)

   (c) (i) List all the possible ways to score a total of 8 (1)

      (ii) What is the probability of scoring 8? Write your answers as a percentage. (1)

Record your score and time here and at the start of the book.

Score [ ] / 25 Time [ ]:[ ]

11+ sample test

1 Write these numbers in order of increasing size. (1)

   0.52   2.52   ⁻3   ⁻2.50   ⁻1   1.25

   −3, −2.50, −1, 0.52, 1.25, 2.52 (4)

2 Round 12.579 to:

   (a) 1 d.p. *12.6*   (b) 2 d.p. *12.58*   (c) 2 s.f.
   (d) the nearest whole number. *13*

3 Here are some number cards. | 9 | 64 | 36 | 5 | 58 | 27 | 44 |

   From these number cards, write down:

   (a) a square number *64* (1)
   (b) a number that can be written as 2 × 2 × 3 × 3 *36* (1)
   (c) a cube number *2 4* (1)
   (d) two numbers that have a product of 45 *9, 5* (1)

4 (a) What is the lowest common multiple of 4, 12 and 18? (1)

   72

   4 = 2 × 2
   12 = 2 × 2 × 3
   18 = 2 × 3 × 3 (1)

   (b) What is the highest common factor of 60 and 80?

   20

   3 × 3 × 4 × 2 =
   9 × 8 = 72

5 I have a pile of 20p coins and a pile of 50p coins. The two piles have the same value.
   What is the least amount of money I could have in total? (1)

   £1

   20 = 10 × 2
   50 = 10 × 5
   10 × 10 =
   100p = £1

6 There were 26 people on a carousel. When the carousel stopped, 18 people got off
   and 13 people got on. How many people were then on the carousel? (1)

7 Calculate: 13 + (⁻27) + 4 − (⁻7) *−6* (1)

   13 − 27 + 8
   13 − 19 = −6

8 Use BIDMAS to work out: 12 + (14 − 3) × 2 (1)

   34   *12 + (11) × 2*

9 The toy shop sold 12 computer games costing £7 each, 55 bouncing balls
   costing £2 each and 3 scooters costing £25 each. How much is this in total? (3)

   12   110   22
   7   2
   84   35

   £269

10 Use a formal method to calculate 5438 – 652 = 4786

Show all your working. (1)

$$\begin{array}{r} 5438 \\ 652 \\ \hline 4786 \end{array}$$

11 Use a formal method to calculate 345 × 14 = 4830

Show all your working. (1)

$$\begin{array}{r} 345 \\ 14 \\ \hline 1380 \\ 3450 \\ \hline 4830 \end{array}$$

12 Use dividing by factors to calculate 256 ÷ 32 = 8 (1)

256 = 2 × 2 × 2 × 2 × 2 × 2 × 2 × 2 = 8
32 = 2 × 2 × 2 × 2 × 2

256 ÷ 32 = 8

13 Use long division to calculate 3012 ÷ 24 (There may be a remainder.) (1)

$$24\overline{)3012}$$ gives 0125

3012 ÷ 24 = 125 r 12

24
48
72
96
120
142

14 £6345 is to be divided equally between two charities. How much will each charity receive? (1)

£3172.50

15 Which one of these multiplications has the largest result? (2)

A: 28 × 4   B: 24 × 8   C: 48 × 2   D: 42 × 8   E: 84 × 2

handwritten: ×2 (24×8) ÷2 = 48×4 ; ×2 (42×8) ÷2 = 84×4 ; D × 4 circled

16 Write these proper fractions in their lowest terms. (3)

(a) $\frac{4}{6}$ *= $\frac{2}{3}$*

(b) $\frac{12}{36}$ *= $\frac{1}{3}$*

(c) $\frac{3}{6}$ *= $\frac{1}{2}$*

17 Write these improper fractions as mixed numbers in their simplest form. (3)

(a) $\frac{6}{4}$ *= $1\frac{1}{2}$*

(b) $\frac{14}{3}$ *= $4\frac{2}{3}$*

(c) $\frac{37}{5}$ *= $7\frac{2}{5}$*

18 There are 16 sweets in a bag. 12 of them are red. What fraction of the sweets is this? Give your answer in its lowest terms. (1)

$\frac{12}{16} = \frac{3}{4}$

19 What is $\frac{5}{12} + 2\frac{5}{6}$? Write your answer as a fraction in its lowest terms. (1)

$\frac{5}{12} + \frac{10}{12} + 2$; $2\frac{15}{12} = 2 + \frac{15}{12}$; $\frac{15}{12} = \frac{5}{4} = 1\frac{1}{4}$; $1\frac{1}{4} + 2 = 3\frac{1}{4}$ (circled)

20 Write the next two fractions in this sequence. Write the fractions in their lowest terms. (1)

$\frac{1}{4}$, 1, $1\frac{3}{4}$, ..., ...   *$2\frac{1}{2}$ $3\frac{1}{4}$*

21 Write down the ratio of purple squares to green squares in this pattern. (1)

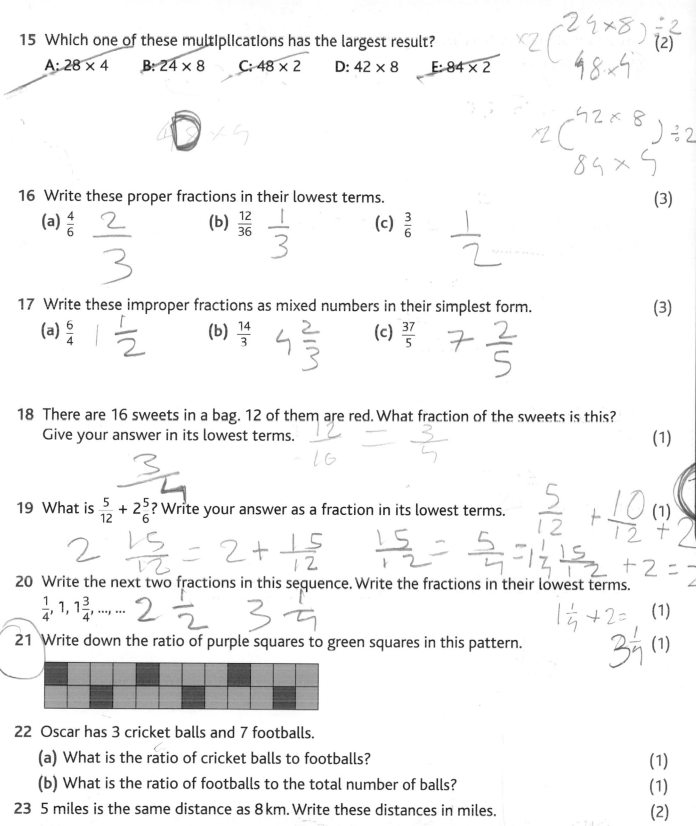

22 Oscar has 3 cricket balls and 7 footballs.

(a) What is the ratio of cricket balls to footballs? (1)

(b) What is the ratio of footballs to the total number of balls? (1)

23 5 miles is the same distance as 8 km. Write these distances in miles. (2)

(a) 16 km   (b) 40 km

24 A model boat is made to a scale of 1 : 50

(a) If the model is 36 cm long, what is the length, in centimetres, of the real boat? (1)

(b) If the real boat is 525 cm wide, how wide is the model? (1)

25 Calculate: (2)

(a) £3.45 × 1000   (b) 23.45 ÷ 100

£3450 *✓ till here*

26 Max is saving up to buy a remote-controlled helicopter that costs £68

If he saves £8 a month, for how many months will he need to save to buy the helicopter? (1)

9 months

27 Write these decimals as fractions in their lowest terms.

(a) 2.25 $2\frac{1}{4}$ (1)

(b) 14.003 $14\frac{3}{1000}$ (1)

(c) 0.0020 $\frac{1}{500}$ (1)

28 How many of these fractions are equivalent to 0.75? (4)

$\frac{3}{4}$ $\frac{15}{20}$ $\frac{25}{75}$ $\frac{75}{100}$ $\frac{27}{36}$ $\frac{9}{12}$ $\frac{12}{48}$ $\frac{3}{4}$ $\frac{3}{4}$

A: 7   **B:** 6   **C: 5**   **D:** 4   **E:** 3   **F:** 2   **G:** 1

5

29 Complete the table below to show equivalent fractions, decimals and percentages. (5)

Fraction (in simplest form)	$\frac{2}{5}$	$1\frac{1}{4}$	$\frac{4}{25}$	$\frac{3}{5}$	$2\frac{1}{4}$
Decimal	0.4	1.25	0.16	0.6	2.25
Percentage	40%	125%	16%	60%	225%

$\frac{16}{100}$ $\frac{4}{25}$

30 Write these numbers in order, largest first. (2)

0.208 $\frac{1}{5}$ 22% $\frac{1}{4}$ ~~0.208~~

0.2 0.22   0.25

$\frac{1}{4}, 22\%, 0.208, \frac{1}{5}$

31 Calculate $\frac{2}{5}$ of £215 (1)

$\frac{2}{5}$ · $\frac{215}{1}$ 86

£86

32 If $\frac{1}{11}$ of a total amount of money is £132, what is the original amount? (1)

£1452 $\frac{132}{11}$ $\frac{132}{132}$ 1452

33 Archie used 10 g of cocoa in his hot chocolate. This was $\frac{5}{7}$ of his total supply.
What was the mass, in grams, of his total supply? (1)

$2g = \frac{1}{7}$

$14g = \frac{7}{7}$ **14 g**   *till here ✓*

34 What is 45% of 40 kg? $\frac{9}{2|0}$ $\frac{20}{1}$ 18 (1)

18 kg

35 A kitchen shop is having a 20% off sale. The full cost of a saucepan is £100.
How much is it in the sale? £80 (1)

36 A pair of shoes cost a shoe shop £25. They were sold at a loss of 16%. Calculate the selling price. (1)
£25 25 $\frac{21}{25}$ £21

37 Give your answers in their lowest terms. (2)
(a) $\frac{2}{5} \times \frac{1}{8}$ $\frac{2}{40} = \frac{1}{20}$ (b) $4\frac{1}{2} \times \frac{1}{4}$

$\frac{1}{20}$

38 Give any improper fractions as mixed numbers in their lowest terms. (2)
(a) $6 \div \frac{2}{5}$ $\frac{36}{1} \times \frac{5}{2} = 15$ (b) $\frac{3}{4} \div \frac{6}{19}$ $\frac{3}{4} \times \frac{19}{6} \frac{19}{8}$

15 $\frac{19}{8}$

39 $\frac{2}{3}$ litre of water is shared equally between 6 children. What fraction of a litre does each child get? (1)
$\frac{2}{3} \div \frac{1}{6} \frac{1}{3}$ $\frac{1}{9}$ of a litre

40 To make 25 pancakes, I need $1\frac{1}{2}$ cups of flour. How many cups of flour will I need to make 15 pancakes? (1)

2.5 : 1.5
5 : 0.3

0.9 cups
of flour or $\frac{9}{10}$
cups
of flour

41 These measuring jugs are empty. 15.0.9
Jug A has a scale up to 1 litre and jug B
has a scale up to 500 millilitres.

(a) How many millilitres does one small division represent on: (2)
(i) jug A 50ml (ii) jug B? 25ml
(b) 450 ml of water is poured into each jug. Draw a line on each jug to show the water level. (2)

42 A ladder is 4.25 metres long. Write this measurement in: (2)
(a) centimetres 425 (b) millimetres. 4250

43 Write $1\frac{3}{4}$ hours in hours and minutes. (1)
1 hour 45 min

44 I drove 70 miles in 2 hours.
What was my average speed on this journey? (1)

45 I cycle for 3 hours at a constant speed of 10 km/h.
What distance do I travel? (1)

46 Which of these quadrilaterals are congruent? (1)

47

(a) What type of shapes are these? (1)

(b) Name shape D. (1)

(c) How many lines of symmetry does shape B have? (1)

(d) How many lines of symmetry does shape D have? (1)

(e) What angle is made where the diagonals of shape A cross? (1)

(f) What is the order of rotational symmetry of shape B? (1)

48 Reflect the shape in the line *AB*. (1)

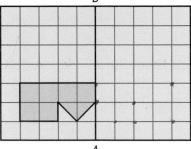

49 *ABCD* is a right-angled trapezium.

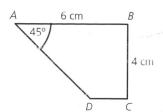

(a) Calculate the size of these angles. (3)

    (i) ∠*B*   **(ii)** ∠*C*   **(iii)** ∠*D*

(b) Draw *ABCD* accurately and measure the lines *CD* and *AD*. (4)

50 (a) How many lines of symmetry does a regular hexagon have? (1)
    (b) What is the order of rotational symmetry of a regular hexagon? (1)

51 What is the radius of a circle with a diameter of 18 cm? (1)

52 A farmer buys an L-shaped field.

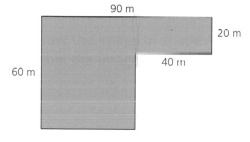

(a) What is the perimeter of the field? Work out any unknown lengths first. (1)

(b) Fencing costs £2.50 per metre. How much will it cost the fence the field? (1)

(c) What is the area of the field? (2)

53 Triangle *ABC* has base *AB* = 12 cm and a height of 3 cm. Calculate its area. (1)

54 A triangle has area 121 m² and a base of 11 m. What is its perpendicular height? (1)

65 Fill in the missing operation(s) (+, −, ×, ÷) in these statements.

(a) 20 ☐ 3 = 60   (1)

(b) 5 ☐ 3 = 20 ☐ 5   (1)

66 The cost of two cupcakes is £1.60. The cost of one brownie and one cupcake is £2.00
What is the cost of two brownies?   (1)

67 Complete this function machine.   (3)

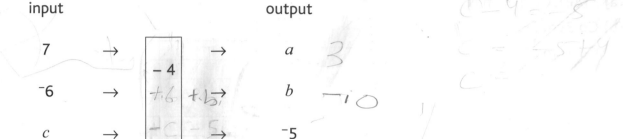

68 Complete this function machine.   (4)

A = −2   B = ×3   a = 36   b = −6

69 Write the next two numbers in this sequence.   (1)

76 70 64 58 ___ ___

70 Write the missing terms in this sequence.   (1)

−0.5, ___, 0, 0.25, ___, 0.75

71 Two numbers have a product of 36 and a sum of 15. What are the two numbers?   (1)

___ and ___

72 India thought of a number. She multiplied it by 12 and then added 4 to the result.

(a) Write an expression to represent her final result.   (1)

(b) If India thought of the number 3, what was the final result?   (1)

73 Write the missing number in this statement.   9 + ☐ = 27   (1)

74 What is the value of the letter a in this equation?   4(a − 2) = 48   (1)

a =

75 Rosa thought of a number, added 6 and then multiplied the result by 7
Her final result was 63 (1)

What number did Rosa think of?

76 (a) Draw the next two patterns in the sequence. (2)

  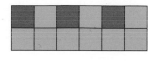

P		P		P		P			P		P		P		P		R

(b) Compete the table for this sequence. (3)

Pattern number	Purple squares	Green squares	Total squares
1	1	3	4
2	2	6	8
3	3	9	12
4	4	12	16
5	5	15	20
6	6	18	24

(c) What is the formula for the total number of green squares in the nth pattern? (1)

77 The table shows the results of a survey about pets owned by pupils in year 6 (4)

Pets owned	Number of pupils
Only dogs	JHT IIII
Only cats	JHT I
Dogs and cats	JHT JHT III
No dogs or cats	JHT

(a) Write the results in the correct regions of this Carroll diagram.

	Cats	No cats
Dogs	13	9
No dogs	6	5

(b) Represent this data in a Venn diagram with two sets (Dog owners and Cat owners).

78 The pictogram shows the spelling test scores for a group of pupils.

■ Marks in a spelling test out of 20

Archie	● ● ● ✓ ● ● ● ● ● ◖
Belinda	● ● ● ● ◖
Charlie	● ● ●
David	●
Ellie	● ● ● ● ● ● ◖

Key: ● represents 2 marks

(a) Who scored the highest mark? (1)

(b) What is the difference between the highest and lowest scores? (1)

(c) How many pupils scored more than half marks? (1)

79 Sebastian counted the number of parents watching rugby matches in the Autumn term. His results are shown below.

Match	1	2	3	4	5	6	7
Number of parents	24	11	8	3	15	3	16

(a) Draw a bar chart to display the results. (3)

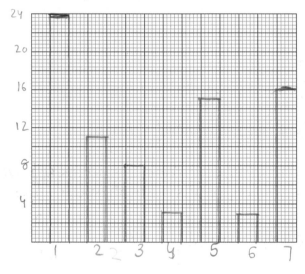

(b) What was the range? (1)

(c) What was the median number of parents watching? (1)

80 The school hockey team played 15 matches this term. The numbers of goals they scored in each match are listed below.

4 0 1 4 3 3 2 0 5 4 3 5 0 4 4

(a) Complete the frequency table. (3)

Goals scored in a match	Tally	Frequency	Fraction	Angle
0	III	3	$\frac{1}{5}$	72°
1	I	1	$\frac{1}{15}$	24°
2	I	1	$\frac{1}{15}$	24°
3	III	3	$\frac{1}{5}$	72°
4	IHI	5	$\frac{3}{?}$	120°
5	II	2	$\frac{2}{15}$	48°
	Total			

(b) Draw a frequency diagram for this data. (2)

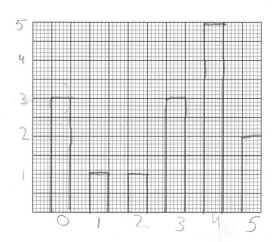

(c) Calculate the angle for each sector and draw a pie chart to display the scores. Add two extra columns to the table: one for 'Fraction' and one for the 'Angle of pie chart'. Use the space above. (2)

81 (a) List all the possible outcomes when an ordinary die is rolled. (1)

(b) What fraction of these outcomes:

  (i) are odd numbers (1)

  (ii) is a 3 or a 4? (1)

(c) If you rolled the die 100 times, how many times might you expect to score an even number? (1)

82 Harry has a coin and a die. He tosses the coin and rolls the die at the same time.

(a) Draw a table to show all the possible outcomes. (1)

(b) How many possible outcomes are there in total? (1)

(c) What is the probability of getting a Head and an even number? (1)

(d) What is the probability of getting a Tail? (1)

83 For this set of numbers, write down: (5)

15 17 14 15 16 12 10 13 14 19 14 9

(a) the numbers in order, smallest first

(b) the range　　　(c) the mode　　　(d) the median　　　(e) the mean.

84 Water was heated and left to cool down. The temperature of the water was recorded every 60 seconds. (3)

Time (minutes)	0	1	2	3	4	5	6	7	8	9	10
Temperature (°C)	90	84	78	69	65	60	52	46	38	26	19

(a) Draw a set of axes on the grid below. Label the horizontal axis 'Time (minutes)' and mark it in intervals of 1 minute. Label the vertical axis 'Temperature °C' and mark it in intervals of 10 °C. Plot the recorded temperatures on the axes with a (+).

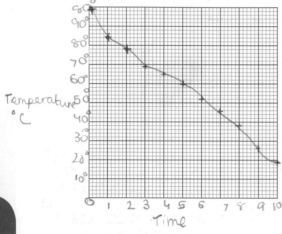

(b) Join the points with a smooth curve.

(c) What was the temperature of the water after 6.5 minutes?

85 This is a conversion graph between US dollars and UK pounds.

(a) What does one small square represent on the vertical axis?

(b) What is $6 worth in pounds?

(c) What is £9 worth in dollars?

■ Conversion between pounds (£) and dollars ($) (3)

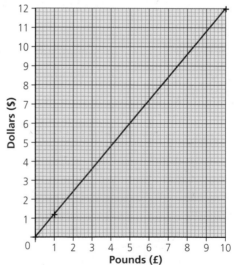

Score ☐ / 200　Time ☐ : ☐

Answers

Chapter 1

A: Place value and ordering

1   732 612   (1)

2   (a) twenty-five thousand, eight hundred and seventy-six   (1)
   (b) two million, four hundred and seventy-three thousand, nine hundred and fifty-three   (1)
   (c) four hundred million, one hundred and thirty-five thousand, seven hundred and forty-eight   (1)

3   (a) fifty (1)   (c) five tenths (1)
   (b) no hundred thousands (1)   (d) eight hundredths (1)

4   (a) 1769 1796 1967 1976   (1)
   (b) 65 ⁻40 ⁻65   (1)

5   (a) 2015 < 2051 (1)   (c) 350 999 > 305 999 (1)
   (b) 145 606 < 154 606 (1)

6   (a) 3000 (3 thousands)   (b) 1000 times greater (2)

7   43.8 79.81 0.85   (1)

8   5.34 4.53 4.35 3.54 3.45   (1)

9

(1)

B: Rounding and estimating

1   (a) 700 (1)   (b) 1400 (1)   (c) 452 800 (1)

2   (a) 1560 (1)   (b) 27 460 (1)   (c) 413 120 (1)

3   (a) 8 (1)   (b) 10 (1)   (c) 6 (1)

4   (a) (i) 7.2 (1) (ii) 7.25 (1) (iii) 7 (1) (iv) 7.2 (1)
   (b) (i) 3.5 (1) (ii) 3.48 (1) (iii) 3 (1) (iv) 3.5 (1)
   (c) (i) 10.6 (1) (ii) 10.57 (1) (iii) 11 (1) (iv) 11 (1)
   (d) (i) 43.5 (1) (ii) 43.48 (1) (iii) 43 (1) (iv) 40 (1)

5   (a) 25 000 + 54 000 = 79 000   (2)
   (b) 730 000 − 450 000 = 280 000   (2)
   (c) 600 × 30 = 18 000   (2)
   (d) 900 ÷ 10 = 90   (2)

6   (a) 200 m (8800 − 8600)   (2)
   (b) 20 m (8610 − 8590)   (2)
   (c) 8700 m   (2)

C: Index numbers, roots, factors and multiples

1   (a) 64 (1)   (b) 5 (1)   (c) 3 (1)

2   (a) 1 × 24 2 × 12 3 × 8 4 × 6   (1)
   (b) 1 × 36 2 × 18 3 × 12 4 × 9 6 × 6   (1)
   (c) 1 × 45 3 × 15 5 × 9   (1)

3   (a) 729 > 36 (1)   (c) 5 = 5 (1)
   (b) 80 < 81 (1)   (d) 8 > 4 (1)

4   (a)

1   2   4   17   34   68   (1)
   (b) 1 × 68 2 × 34 4 × 17   (1)

5   9   (1)

---

6   $a = 4$ and $b = 3$ (or $a = 8$ and $b = 2$)   (1)

7   (a) 35 (1) (c) 4 (1) (e) 15 (1) (g) 27 (1)
   (b) 5 (1) (d) = 24 (1) (f) 4, 12 (1)

D: Prime numbers, prime factors, HCF and LCM

1   B: 4   (1)

2   (a) $2^3 × 3$ (1)   (b) $2 × 13$ (1)   (c) 3^3 (1)

3   (a) 4 (factors of 12 are 1, 2, 3, **4**, 6, 12; factors of 20 are 1, 2, **4**, 5, 10, 12)   (1)
   (b) 36 (multiples of 12 are 12, 24, **36**, 48, …; multiples of 18 are 18, **36**, …)   (1)

4

(1)

5   (a) $3^2 × 5^2$ (1)   (b) $2 × 3^2 × 13$ (1)   (c) $2^5 × 3 × 5$ (1)

6   (a) 50 (factors of 100 are 1, 2, 4, 5, 10, 20, 25, **50**, 100; factors of 150 are 1, 2, 3, 5, 6, 10, 15 ,25, 30, **50**, 75, 150)   (2)
   (b) 45 (multiples of 5 are 5, 10, 15, 20, 25, 30, 35, 40, **45**, 50, …; multiples of 9 are 9, 18, 27, 36, **45**, …; multiples of 15 are 15, 30, **45**, …)   (3)

7   $2^3 × 5 × 17$ (1)

8   24 seconds (multiples of 8 are 8, 16, **24**, 32, …; multiples of 12 are 12, **24**, …)   (1)

9   60p (multiples of 5 are 5, 10, 15, 20, 25, 30, 35, 40, 45, 50, 55, **60**, 65, …; multiples of 20 are 20, 40, **60**, …)   (1)

Test 1

1   D: 3.45 (0.345, 3.045, **3.45**, 340, 3 405)   (1)

2   D: 8   (1)

3   (a) 100 times   (1)   (d) 1000 times   (1)
   (b) 1000 times   (1)   (e) 100 000 times (1)
   (c) 10 000 times   (1)

4   (a) 89 (1)   (b) 2 (1)

5   (a) 1 480 000 + 3 410 000 = 4 890 000   (1)
   (b) 5 567 000 − 2 546 000 = 3 021 000   (1)

6   (a) 18 ($2 × 3 × 3$) and 32 ($2 × 2 × 2 × 2 × 2$ or 2^5) (1)
   (b) 2   (1)

7   3 (48 is $2 × 2 × 2 × 2 × 3$; 75 is **3** $× 5 × 5$; 120 is $2 × 2 × 2 × $**3**$ × 5$)   (2)

8   168 (multiples of 24 are 24, 48, 72, 96, 120, 144, **168**, …; multiples of 56 are 56, 112, **168**, 224, …)   (2)

9   (a) (i) 29.09 35.66 43.25 45.04   (1)
      (ii) 36 43 45 30   (2)
      (iii) 35.7 43.3 45.0 29.9   (2)
      (iv) 40 40 50 30   (2)
   (b) 6 cm is the longest length of string ($24 = 2 × 2 × $**2**$ × 3$; $18 = $**2**$ × 3 × 3$)   (2)
   (c) 3 p.m. (multiples of 2 are 2, 4, **6**, 8, …; multiples 3 are 3, **6**, 9, …)   (2)

Chapter 2 Calculations

A: Mental strategies

1. (a) 18 (1) (d) ⁻18 (1) (g) 18 (1) (j) $-\frac{1}{2}$ (1)
   (b) 18 (1) (e) ⁻18 (1) (h) 6 (1)
   (c) ⁻6 (1) (f) ⁻6 (1) (i) 72 (1)
2. (a) 149 (1) (c) 112 (1) (e) 32 (1) (g) 35 (1)
   (b) 900 (1) (d) 299 (1) (f) 74 (1) (h) 85 (1)
3. 34 (2)
4. (a) 160 (1) (b) 1.6 (1) (c) 1.6 (1) (d) 0.8 (1)

B: Negative numbers

1. (a) 2 (1) (c) 0 (1) (e) ⁻6 (1)
   (b) ⁻4 (1) (d) ⁻2 (1) (f) ⁻13 (1)
2. (a) 3 (1) (c) ⁻10 (1) (e) 10 (1)
   (b) ⁻1 (1) (d) 10 (1) (f) 2 (1)
3. (a) 1 (1) (c) 7 (1) (e) ⁻10 (1)
   (b) ⁻10 (1) (d) ⁻14 (1) (f) ⁻57 (1)
4. (a) 20 (1) (c) ⁻25 (1) (e) 1 (1)
   (b) ⁻3 (1) (d) 20 (1)
5. (a) 5 (1) (b) ⁻4 (1) (c) ⁻11 (1)
6. (a) 22, 30 (1) (b) ⁻1.2, ⁻0.6 (1)
7. (a) 6 °F (1) (b) 29 °F (1)

C: Order of operations

1. (a) 6 (1) (c) 6 (1) (e) 36 (1) (g) 141 (1)
   (b) 28 (1) (d) 13 (1) (f) 26 (1) (h) 2 (1)
2. (a) 165 (1) (b) 15 (1) (c) 10 (1) (d) 23 (1)
3. In total, 67 pots of paint have been ordered. (5)
4. E: 58 (A: 6 × 9 = 54 B: 7 × 8 = 56 C: 110
   2 = 55 D 280 ÷ 5 = 56 E: 29 × 2 = 58) (5)
5. (a) 20 × 10 → 200, so approximately 200 poems
   were given out. (1)
   (b) 252 poems were given out. (1)
6. 10 and 15 (10 + 15 = 25 and 10 × 15 = 150) (2)
7. £350 was taken by the three stalls (25 × £4 = £100
   65 × £2 = £130 and 20 × £6 = £120) (4)

D: Formal methods – addition and subtraction

1. (a) (1)

H	T	U
	1	4	3
+		3	7
	1	8	0
		1	

(b) (1)

H	T	U
⁰⧸	¹⁰⧸	¹1
−	0	3	4
	0	7	7

(c) (1)

T	U	•	t
	5	•	7
+	6	•	9
1	2	•	6
1	1		

(d) (1)

T	U	•	t	h
1	3	•	⁴⧸	¹0
−	2	•	4	8
1	1	•	0	2

(e) (1)

Th	H	T	U
	1	3	4	5
+	5	6	8	7
	7	0	3	2
	1	1	1	

(f) (1)

Th	H	T	U
	6	7	9	8
−	2	3	4	2
	4	4	5	6

2. (a) (1)

T	U	•	t
⁰⧸	¹¹⧸	•	¹3
−		5	•	6
		6	•	7

(b) (1)

T	U	•	t
2	²⧸	•	¹2
−	1	1	•	4
	1	¹1	•	8

(c) (1)

TTh	Th	H	T	U
8	9	5	6	8
	5	6	4	3
		2	1	1
+			5	6
9	5	4	7	8
1	1	1	1	

(d) (1)

TTh	Th	H	T	U
⧸	¹⁷⧸	⁹¹⧸	⁹¹⧸	¹1
−		9	3	9	8
		8	6	0	3

3. 2642 (1)
4. (a) 40 018 (1) (b) 24 889 (1)

E: Short multiplication and multiplication by multiples of 10

1. (a) (1)

H	T	U
	1	9
×		8
1	5	2
	7	

(b) (1)

H	T	U
	5	6
×		6
3	3	6
	3	

(c) (1)

Th	H	T	U
	6	7	1
×			4
2	6	8	4
	2		

(d) (1)

TTh	Th	H	T	U
		1	7	9	5
×					9
1	6	1	5	5
		⁷	⁸	⁴

(e) 256 100 (1) (f) 256 700 (1)

2. (a) (1)

Th	H	T	U
	2	4	5
×			7
1	7	1	5
	3	3	

(b) (1)

HTh	TTh	Th	H	T	U
		1	9	4	3
×				6	0̸
	1	1	6	5	8	0
		₅	₂	₁	

(c) (1)

TM	M	HTh	TTh	Th	H	T	U
				4	2	2	2
×					8	0̸	0̸	0̸
	3	3	7	7	6	0	0	0
			₁	₁	₁		

3 £43.75 (There are 2 figures after the decimal point in the questions, so there must be 2 figures after the decimal point in the answer.)

Th	H	T	U
	8	7	5
×			5
4	3	7	5
	₃	₂	

4 D: 42 × 5 = 210 (1)

5 (a) (i) smallest 78 × 6 ≈ 80 × 5 = 400 (5)

(ii) largest 86 × 7 ≈ 85 × 10 = 850 (3)

(b) (3)

H	T	U
	6	7
×		8
5	3	6
	₅	

H	T	U
	8	6
×		7
6	0	2
	₄	

H	T	U
	7	8
×		6
4	6	8
	₄	

6 (1)

TM	M	HTh	TTh	Th	H	T	U
				8	6	3	0
×					7	0̸	0̸	0̸
6	0	4	1	0	0	0	0
		₄	₂				

7 The airport bus carries 6240 people in a week. (1)

8 The chef will need to order 112 000 g of flour. (2)

F: Multiplication by a two-digit number

1 218 × 16 = 3488 (1) 302 × 28 = 8456 (1)

193 × 32 = 6176 (1) 412 × 14 = 5768 (1)

2 (a) 45 × 3 × 2 × 2 = 540 (2)

(b) 122 × 3 × 5 = 1830 (2)

3 (a) (3)

TTh	Th	H	T	U
		5	1	2
×			2	0̸
1	0	2	4	0
		₅		

TTh	Th	H	T	U
		5	1	2
×				8
	4	0	9	6
		₁		

TTh	Th	H	T	U
	1	0	2	4	0
+		4	0	9	6
1	4	3	3	6
		₁		

(b) (3)

TTh	Th	H	T	U
		1	6	4	0
×				3	0̸
4	9	2	0	0
	₁	₁		

TTh	Th	H	T	U
		1	6	4	0
×				7
1	1	4	8	0
		₄	₂	

TTh	Th	H	T	U
	4	9	2	0	0
+	1	1	4	8	0
6	0	6	8	0
	₁			

4 (a) (3)

	Th	H	T	U
		2	4	6
×			2	1
		2	4	6	×1
+	4	9¹	2	0	×20
	5	1	6	6
	₁			

(b) (3)

TTh	Th	H	T	U
		8	9	5
×			4	9
	8	0⁸	5⁴	5	×9
+	3	5	8	0	0	×40
4	3³	8²	5	5
₁				

5 888 (2)

6 14 336 (2)

7 Total cost £8014 (1) (£2758 adults (1), £5256 children (1))

G: Short division, division by factors and multiples of 10

1 (a) 218 (1) (c) 244 r 1 (1)

(b) 617 (1) (d) 982 r 7 (1)

2 (a) 11 (1) (c) 273 (1) (e) 90 (1)

(b) 68 (1) (d) 68 (1) (f) 40 (1)

3 (a) 120 (1) (b) 1.2 (1) (c) 1.2 (1) (d) 0.6 (1)

4 1782 full buckets (2)

5 96 feet (2)

H: Long division

1 (a) 145 (1) (c) 343 (1) (e) 37 (1)

(b) 331 (1) (d) 36 (1) (f) 18 (1)

2 (a) $68\frac{1}{3}$ or 68 r 2 (1) (d) 122 (1)

(b) $68\frac{1}{8}$ or 68 r 4 (1) (e) 140 r 21 or $140\frac{1}{2}$ (1)

(c) 109 (1) (f) 167 r 12 or $167\frac{1}{3}$ (1)

3 20 (1)

4 125 cupcakes (1)

5 18 buses (2)

6 Each charity will receive £137 (£3 remaining) (1)

Test 2

1 300 (1)

2 79 (1)

3 12.5 (1)

4 11.01 (1)

5 398 (1)

6 4960 (1)

7 9 (1)

8 11.9 (1)

9 (a) 203 (1) (b) 20 (1) (c) 6384 (1)

10 (a) 409 (1) (b) 286 (1) (c) 258 r 3 or $258\frac{1}{3}$ (1)

11 B: 4 (2)

12 (a) 144 (1) (b) 45 (1)

13 850 minutes (1)

14 (a) £5.71 each person (1)   (b) 3p left (1)
15 They can feed 250 cats.   (2)
16 56 rows   (2)

Chapter 3 Fractions, proportions and percentages

A: Equivalent fractions, improper fractions and mixed numbers

1    (1)

2 (a) Yes, $\frac{1}{2}$ (1)   (b) Yes, $\frac{3}{4}$ (1)   (c) Yes, $\frac{3}{4}$ (1)   (d) No (1)

3 (a) $\frac{1}{3}$ (1)   (b) $\frac{1}{2}$ (1)   (c) $\frac{8}{9}$ (1)   (d) $\frac{13}{36}$ (1)

4 (a) $\frac{1}{4} = \frac{4}{16}$ (1)   (c) $\frac{8}{9} = \frac{40}{45}$ (1)   (e) $\frac{5}{6} = \frac{20}{24}$ (1)

   (b) $\frac{7}{15} = \frac{14}{30}$ (1)   (d) $\frac{2}{3} = \frac{10}{15}$ (1)   (f) $\frac{3}{5} = \frac{15}{25}$ (1)

5 (a) $\frac{1}{2}$ (1)   (b) $\frac{3}{5}$ (1)   (c) $\frac{3}{4}$ (1)

6 (a) $2\frac{1}{4}$ (1)   (b) $2\frac{2}{5}$ (1)   (c) $6\frac{1}{3}$ (1)

7 $\frac{9}{12} = \frac{3}{4}$ (2)

8 $\frac{10}{32}$ are boys, so $\frac{22}{32} = \frac{11}{16}$ are girls.   (2)

B: Ordering, adding and subtracting fractions; fraction sequences

1 (a) $\frac{4}{7}$ $\frac{1}{2}$ (1)   (b) $\frac{7}{9}$ $\frac{8}{12}$ (1)   (c) $\frac{16}{6}$ $1\frac{2}{3}$ $1\frac{1}{4}$ (1)

2 (a) $\frac{3}{5}$ (1)   (b) $1\frac{3}{4}$ (1)   (c) $\frac{1}{2}$ (1)   (d) $3\frac{5}{21}$ (1)

3 (a) $10\frac{11}{60}$ (1)   (b) $2\frac{3}{10}$ (1)

4 (a) $2\frac{1}{30}$ (1)   (b) $2\frac{1}{15}$ (1)

5 (a) $2\frac{1}{3}$ 3 (2)   (b) $1\frac{1}{4}$ $1\frac{7}{16}$ (2)

6 $11\frac{11}{12}$ miles in total   (2)

C: Ratio

1 (a) 3:4 (1)   (b) 2:9 (1)   (c) 3:8 (1)
2 (a) 1:1 (1)   (b) 1:2 (1)
3 (a) 7:3 (1)   (b) 7:10 (1)
4 (a) 4:1 (1)   (c) 25 kg of sugar (1)
   (b) 128 kg of strawberries (1)
5 (a) 8 are mint (1)   (c) 30 chocolate and 24 mint (1)
   (b) 18 in a small box (1)
6 (a) 9, 6 (1)   (b) 27 biscuits (1)   (c) 2:1:2 (2)

D: Proportion, equivalent measures, ratio and scale

1 (a) 16 km (1)   (b) 104 km (1)   (c) 256 km (1)
2 (a) 11 lb (1)   (b) 33 lb (1)
3 (a) 9 eggs (1)   (b) 30 pancakes (1)   (c) 460 g flour (1)
4 $\frac{2}{5}$ are girls   (1)
5 (a) 15:9 (1)   (c) 180 g chemical Y (1)
   (b) 6 mixtures (1)
6 (a) 1200 cm (1)   (b) 8.5 cm (1)

E: Calculating with money and other decimal calculations

1 (a) £7.02 (1)   (c) £42.06 (1)
   (b) £5.55 (1)   (d) £5.25 (1)
2 2 pizzas £8.50, 3 packets of tortellini £5.70, total £15.80   (3)
3 (a) 1515 (1)   (c) 129 (1)
   (b) 0.002 75 (1)   (d) 2.021 (1)
4 (a) 2.7 (1)   (b) 2.74 (1)   (c) 2.738 (1)
5 (a) £7.07 (1)   (b) £12.93 (1)
6 (a) 789.12 (1)   (b) 0.075 (1)
7 9 months (8.5 rounded up)   (1)
8 10.121   (1)

F: Fractions and decimals

1 (a) 0.67 (1)   (c) 0.11 (1)
   (b) 0.83 (1)   (d) 0.025 (1)
2 (a) $\frac{1}{4}$ (1)   (c) $11\frac{1}{1000}$ (1)

   (b) $1\frac{3}{5}$ (1)   (d) $12\frac{57}{125}$ (1)

3 (a) 0.112 (1)   (b) 0.36 (1)   (c) 0.14 (1)
4 (a) $4\frac{27}{100}$ (1)   (b) $12\frac{1}{250}$ (1)   (c) $\frac{1}{100}$ (1)
5 D: 4   (2)
6 (a) 0.025 (1)   (b) 0.375 (1)
7 (a) $\frac{7}{20}$ (1)   (b) $1\frac{93}{250}$ (1)
8 (a) 0.625 (1)   (b) 0.375 (1)

G: Percentages, decimals and fractions

1 (a) (i) 0.05 (1)   (ii) $\frac{1}{20}$ (1)

   (b) (i) 0.10 (1)   (ii) $\frac{1}{10}$ (1)

   (c) (i) 0.55 (1)   (ii) $\frac{11}{20}$ (1)

2 (a) 10% (1)   (b) 42% (1)   (c) 98% (1)
3 (a) 75% (1)   (b) 40% (1)   (c) 2.5% (1)
4 (a) (i) 0.97 (1)   (ii) $\frac{97}{100}$ (1)

   (b) (i) 1.45 (1)   (ii) $1\frac{9}{20}$ (1)

   (c) (i) 0.605 (1)   (ii) $\frac{121}{200}$ (1)

5 (a) 45% (1)   (b) 176% (1)   (c) 1698% (1)
6 (a) 110% (1)   (b) 216% (1)   (c) 115% (1)
7   (9)

Fraction (in simplest form)	$\frac{2}{5}$	$\frac{1}{4}$	$\frac{7}{20}$	$\frac{4}{5}$	$\frac{7}{10}$
Decimal	0.4	0.25	0.35	0.8	0.7
Percentage	40%	25%	35%	80%	70%

8 (a) (i) 87% (1)   (ii) 88% (1)   (iii) 85% (1)
   (b) English   (1)

H: Recurring decimals; ordering fractions, decimals and percentages

1 (a) $0.\dot{6}$ (1)   (b) $0.\dot{1}$ (1)   (c) $0.2\dot{1}$ (1)

2 (a) (i) 44.44% (2 d.p.) (1)   (ii) 83.33% (2 d.p.) (1)

   (b) (i) $\frac{11}{25}$ (1)   (ii) $\frac{83}{100}$ (1)

3 (a) $0.\dot{0}\dot{9}$ (1)   (c) $0.\dot{2}\dot{7}$ (1)   (e) $0.\dot{4}\dot{5}$ (1)
   (b) $0.\dot{1}\dot{8}$ (1)   (d) $0.\dot{3}\dot{6}$ (1)

4 0.143, 0.286, 0.429, 0.571, 0.714, 0.857   (6)

5 (a) 44% $\frac{47}{100}$ $\frac{12}{25}$ 0.49   (1)

   (b) $\frac{41}{50}$ $\frac{55}{65}$ 85%   0.87   (1)

6   (14)

Fractions	Decimal	Percentage
$\frac{1}{2}$	0.5	50%
$\frac{1}{3}$	0.333...	$33\frac{1}{3}$%
$\frac{2}{3}$	0.666...	66.67%
$\frac{1}{4}$	0.25	25%
$\frac{3}{4}$	0.75	75%
$\frac{1}{5}$	0.20	20%
$\frac{1}{7}$	0.142857...	14.29%

7 (a) $\frac{1}{4}$ 22%   0.208 $\frac{1}{5}$ (1)

   (b) 112% $1\frac{1}{9}$ $1\frac{1}{12}$ 1.02   1.012   (1)

I: Fractions of numbers and quantities

1 (a) 0.5 m (1)   (c) 65 m (1)
   (b) £33.33 (1)   (d) 32.5 km (1)
2 (a) 8 kg (1)   (c) 12.60 litres (1)
   (b) £16.80 (1)   (d) 1.25 km (1)
3 250 kg in each jar   (1)
4 (a) £1.20 (1)   (c) 0.75 km (1)
   (b) 22.8 m (1)   (d) £41.80 (1)
5 (a) £156.25 (1)   (b) 42 litres (1)
6 (a) 48 (1)   (b) 25 (1)
7 750 ml   (1)
8 376.25 km   (1)

J: Finding the original amount

1 8   (1)
2 36   (1)
3 70   (1)
4 27   (1)
5 63   (1)
6 144   (1)
7 33   (1)
8 180   (1)
9 81   (1)
10 9   (1)
11 40   (1)
12 22   (1)

13 16   (1)
14 42   (1)
15 210   (1)
16 £60   (1)
17 5.4 kg   (1)
18 54 kg   (1)

K: Percentage of a quantity and as a fraction

1 English 88%, mathematics 93.3%, science 88%,
   history 86.7%   (4)
2 (a) 6 (1)   (b) 60 (1)   (c) 2.1 (1)
3 82%   (1)
4 Yes, she had improved from 80% to 92%.   (2)
5 (a) 19.5 (1)   (c) 8.4 (1)
   (b) 61.2 (1)   (d) 225 mg (1)
6 (a) 67.8% would vote (answer rounded)   (1)
   (b) 8.3% would not vote   (1)
   (c) 23.9% undecided (answer rounded)   (1)
7 (a) 36% male (1)   (b) 128 females (1)
8 (a) 18 mints (1)   (b) 30 toffees (1)
9 38 25 mm   (1)
10 396 kg   (1)

L: Percentage discounts and charges; profit and loss

1 £31.50   (2)
2 84p   (2)
3 £571.20   (2)
4 33.3%   (2)
5 36%   (2)
6 (a) £24 (1)   (b) £59.16 (1)   (c) £894 (1)
7 A: £46.40, B: £46.50, C: £45.33, D: 30% discount   (4)
8 £97 750   (1)
9 40%   (2)

M: Multiplying with fractions

1 (a) $1\frac{1}{2}$ (1)   (c) $\frac{1}{3}$ (1)   (e) $6\frac{1}{4}$ (1)

   (b) 4 (1)   (d) $4\frac{2}{5}$ (1)   (f) $8\frac{3}{5}$ (1)

2 (a) $\frac{2}{3}$ (1)   (c) $\frac{9}{40}$ (1)   (e) $1\frac{1}{2}$ (1)

   (b) $\frac{2}{27}$ (1)   (d) $\frac{8}{15}$ (1)   (f) $\frac{4}{5}$ (1)

3 (a) $3\frac{1}{3}$ (1)   (c) $1\frac{5}{9}$ (1)   (e) $3\frac{3}{5}$ (1)

   (b) $1\frac{1}{4}$ (1)   (d) $19\frac{1}{2}$ (1)   (f) $24\frac{1}{2}$ (1)

4 (a) $\frac{1}{6}$ (1)   (d) $\frac{2}{9}$ (1)   (g) $1\frac{11}{18}$ (1)

   (b) $\frac{1}{4}$ (1)   (e) $\frac{1}{6}$ (1)   (h) $1\frac{2}{3}$ (1)

   (c) $\frac{3}{14}$ (1)   (f) $1\frac{1}{8}$ (1)   (i) $4\frac{1}{2}$ (1)

N: Dividing by fractions

1 (a) 4 (1)   (c) $1\frac{1}{2}$ (1)   (e) 2 (1)

   (b) 24 (1)   (d) 4 (1)   (f) 40 (1)

2 (a) 3 (1)   (c) $\frac{4}{9}$ (1)   (e) $6\frac{1}{20}$ (1)

   (b) 15 (1)   (d) $1\frac{1}{3}$ (1)   (f) $2\frac{2}{3}$ (1)

3 (a) $4\frac{2}{5}$ (1) (c) $\frac{3}{5}$ (1) (e) $2\frac{1}{4}$ (1)

(b) 18 (1) (d) $1\frac{1}{2}$ (1) (f) $3\frac{1}{7}$ (1)

4 $\frac{1}{20}$ (1)

5 (i) Sheep have $\frac{7}{40}$ of the bucket (1)

(ii) Goats have $\frac{21}{40}$ (1)

Test 3

1 $\frac{3}{4}$ (1)

2 (a) $\frac{3}{12}$ (b) $\frac{8}{20}$ (2)

3 $\frac{1}{4}$ $\frac{2}{5}$ $\frac{1}{2}$ $\frac{2}{3}$ $\frac{3}{4}$ $\frac{5}{6}$ (1)

4 (a) $\frac{11}{4}$ (1) (b) $\frac{39}{20}$ (1)

5 (a) $3\frac{1}{2}$ (1) (b) $5\frac{2}{5}$ (1)

6 (a) $1\frac{1}{6}$ (1) (b) $\frac{9}{10}$ (1)

7 (a) 2500 ml of milk (1)
   (b) 1000 ml of liquid chocolate (1)

8 72 g (1)

9 $\frac{13}{20}$ 0.665 $\frac{2}{3}$ 67% (4)

10 £36 (1)
11 833.33 millilitres (1)
12 16 (2)
13 160 kg (2)
14 maths 74% and science 68% (2)
15 £2.25 (1)
16 £22 (1)

17 (a) $\frac{9}{20}$ (1) (b) $3\frac{7}{10}$ (1)

18 1 cup of flour (2)

Chapter 4: Measures, shape and space

A: Reading number scales

1 B, A, C, D (2)
2 (a) mm (1) (b) km (1)
3 (a) mg (1) (b) tonne (1)
4 13 degrees (1)
5 (a) 45 000 grams (1) (c) 7 stones and 1 pound (1)
   (b) 99 pounds (1)
6 (a) 305 centimetres (1)
   (b) 3050 millimetres (1)
   (c) 10 feet (1)
7 6 degrees (1)
8 (a) butter 0.25 kg (1), sugar 0.25 kg (1), flour 0.35 g (1),
      chocolate chips 0.55 kg (1), eggs 0.25 kg (1)
   (b) 1.65 kg (1) (c) 1.40 kg (rounded to 2 d.p.) (1)

B: Working with time

1 120 minutes (1)
2 2880 minutes (1)
3 (a) 20:40 (1) (b) 23:59 (1) (c) 12.16 (1)

4 (a) $\frac{1}{3}$ (1) (b) $1\frac{3}{4}$ (1)
5 47 days (5 in Sept, 31 Oct and 11 Nov) (1)
6 (a) 15 minutes past 2 (or quarter past 2) (1)
   (b) 25 minutes to 8 (1)
   (c) 6 minutes to midnight (1)
7 22:11 in Sydney (1)
8 (a) 15 mins (1)
   (b) 3 hrs and 40 mins (1)
   (c) 2 hrs and 25 mins (1)
9 73 days (1)
10 (a) 2h 51m (1) (b) 15:29 (1) (c) 4h 28m (1)
11 $3\frac{2}{3}$ (1)
12 4 hrs and 50 mins (1)

C: Speed, distance and time

1 (a) Car (1) (b) Walk (1) (c) Bicycle (1) (d) Car (1)
2 50 mph (1)
3 120 mph (1)
4 3 km (1)
5 (a) I will arrive first. (1)
   (b) We arrive at the same time. (1)
6 10 km/h (1)
7 2 miles (1)
8 20 mins (1)
9 The red car (1)
10 100 kmh (1)
11 Jasper (2)
12 340 km (1)
13 14:30 (2)

D: Congruent shapes and similar shapes

1 (a) D and E (1) (b) C (1)
2 A and C (1)
3 (a) C and E (1) (b) D (1)
4 (a) 1:2 (1)
   (b) (i) 5 cm (1) (ii) 4 cm (1)
5 A and C (2)
6 (a) (1)

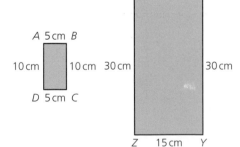

(b) 30 cm (1)
(c) 1:3 (1) (d) Scale factor 3 (1)

E: Plane shapes

1 (a) (1)

(b) Any 4-sided shape that is not a square (1)

2 (a) (3) 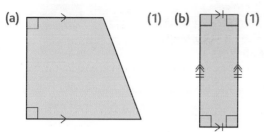 (b) (3)

(c) 1 line of symmetry (1)

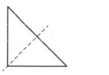

3 (a) (1) (b) (1)

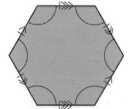

3 B: 5 (4)

4 (a) (i) Kite (1)
  (ii) 1 line of symmetry (1)

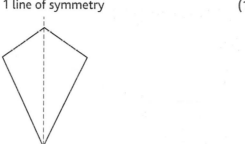

4 (a) 6 equal sides, 6 equal angles, 3 pairs of parallel lines (4)

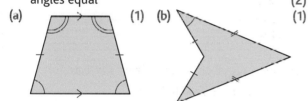

(iii) no rotational symmetry (1)
(b) (i) Parallelogram (1)
   (ii) no symmetry (1)

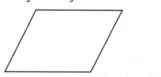

(b) Any 6-sided shape with not all sides and angles equal (2)

(iii) rotational symmetry order 2 (1)
(c) (i) Star (1)
   (ii) 4 lines of symmetry (1)

5 (a) (1) (b) (1)

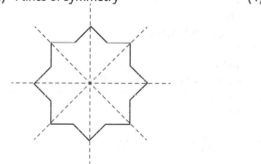

F: Symmetry

1 (a) (1)

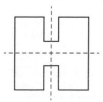

(iii) rotational symmetry order 8 (1)

G: Properties of quadrilaterals

(b) (1)

1 (a) diagonals are equal, one diagonal bisects the other, the diagonals meet at right angles, both diagonal are lines of symmetry (3)

(c) (1)

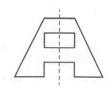

(b) one diagonal bisects the other (3)

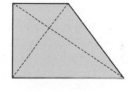

2 (a) 2 lines of symmetry (1) (b) 1 line of symmetry (1)

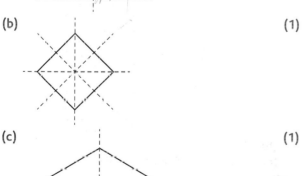

(c) one diagonal is a line of symmetry (3)

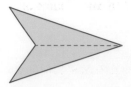

2 Rhombus (2)
3 opposite angles equal, all sides equal length, 2 lines of symmetry, two diagonals which cross at 90°, diagonals not equal, opposite sides parallel (6)

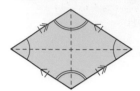

4 3 (1)
5 Square (1)
6 Rectangle (1)
7 Trapezium (1)

H: Angles and lengths in quadrilaterals; drawing quadrilaterals

1 (a) (i) 6 cm (1) (ii) 6 cm (1) (iii) 90° (2)
   (b) 360° (1)
2 (a) (i) 4 cm (1) (ii) 2 cm (1) (iii) 90° (1)
   (b) AC = 4.5 cm and BD = 4.5 cm (3)

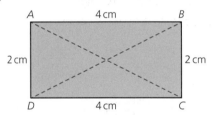

3 (a) (i) 90° (1) (ii) 90° (1) (iii) 127° (1)
   (b) EF = 20 mm and FG = 25 mm

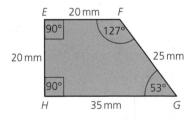
(3)

I: Polygons

1 (a) (i) (1) (ii) (1)

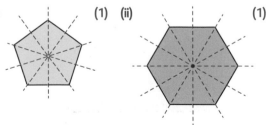

   (b) (i) rotational symmetry order 5 (1)
       (ii) rotational symmetry order 6 (1)
   (c) (i) 5 lines of symmetry (1)
       (ii) 6 lines of symmetry (1)

2
Number of sides n	Name of regular polygon	Size of exterior angle	Size of interior angle
3	Equilateral triangle	120°	60°
4	Square	90°	90°
(6)

3
Number of sides n	Name of regular polygon	Size of exterior angle	Size of interior angle
5	Pentagon	72°	108°
6	Hexagon	60°	120°
7	Heptagon	51.43°	128.57°
8	Octagon	45°	135°
9	Nonagon	40°	140°
(15)

4 (a) 8 lines of symmetry (1)
   (b) rotational symmetry order 8 (1)
5 (a) The interior angle and the exterior angle of a regular polygon add up to **180°** (1)
   (b) The exterior angle of a regular polygon is calculated by the formula $\dfrac{360°}{n}$ (1)
   (c) The interior angle of a regular polygon is equal to **180°** minus $\dfrac{360°}{n}$ (1)

J: Circles

1 8 cm (1)
2 6 cm (1)
3 9 m (1)
4 (a) and (b)

(1)
   (c) 13 cm (1)
5 7 cm (1)
6 12.5 cm (1)
7 (a) Circle with radius of 2 cm drawn (1)
   (b) Any 2 diameters drawn (1)
   (c) *Check measurements* (1)
   (d) 360° (1)
8 (a), (b), (c) (Drawn at 50%) (3)

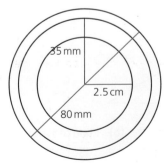

K: Perimeters of squares, rectangles and compound shapes

1 (a) 22 km (1) (b) 160 m (1) (c) 8 cm (1)
2 (a) 124 cm (1) (b) 20 m (1) (c) 220 mm (1)

11 + Mathematics Revision Guide published by Galore Park

3 (a) 22 m (missing lengths 5 m and 4 m) (1)
(b) 240 mm (1) (c) 64 m (1)
4 (a) 130 cm (1) (c) 1.5 m (1)
(b) 150 cm (1) (d) £6.75 (1)

L: Areas of squares, rectangles and compound shapes

1 (a) 49 cm² (1) (b) 25 m² (1) (c) 400 km² (1)
2 (a) 72 mm² (1) (b) 7500 cm² (1) (c) 750 m² (1)
3 (a) 82 cm² (1) (b) 13 m² (1) (c) 17 200 cm² (1)
4 (a) 484 m² (1) (b) 88 m (1) (c) 44 fence panels (1)

M: Areas of triangles and parallelograms

1 (a) 700 cm² (1) (b) 40 000 m² (1) (c) 7.5 km² (1)
2 (a) 12.15 cm² (1) (b) 160 mm² (1) (c) 5250 mm² (1)
3 (a) 48 cm² (1) (b) 17.28 m² (1) (c) 48 in² (1)
4 72 mm² (1)
5 7 cm² (1)

N: Calculating unknown lengths

1 (a) 22 mm (2) (b) 4 km (2) (c) 16 cm (2)
2 6 m (2)
3 50 m (2)
4 11 cm (2)
5 40 m (2)
6 (a) 24 m (2) (b) 30 m (2) (c) 78 m² (2)
7 24 cm (2)
8 3.3 m (2)

O: Three-dimensional shapes

1 (a) 12 (1) (b) 12 (2)
2 (a) *Accurate net drawn*, for example: (50% of actual size): (3)

(b) 62 cm² (3)
3 (a) (12)

Solid shape	Number of faces	Number of vertices	Number of edges
Cube	6	8	12
Cuboid	6	8	12
Square pyramid	5	5	8
Triangular prism	5	6	9

(b) There are 11 possible nets of a cube. Here are three examples: (3)

4 (a) 6 faces (1) (b) 2 (1) (c) 6 (1) (d) 3, 4, 2 and 5 (1)

P: Capacity and volume

1 (a) 40 cm³ (1) (b) 270 m³ (1)
2 343 cm³ (1)
3 420 (1)

4 (a) 5 cm³ (2) (b) 7 cm³ (2) (c) 9 cm³ (2)
5 Cube has larger volume 216 cm³ (volume of cuboid, 192 cm³) (2)
6 1000 m³ (2)
7 (a) 2 cm (1) (b) 5 cm (1) (c) 1 cm (1)
8 2 cm × 3 cm × 2 cm, 6 cm × 2 cm × 1 cm, 4 cm × 3 cm × 1 cm (4)

Q: Angles

1 (a) (i) acute (1) (ii) estimate 30° (1)
(b) (i) obtuse (1) (ii) estimate 110° (1)
(c) (i) reflex (1) (ii) estimate 195° (1)
2 (a) reflex (1) (b) 300° (1)
3 (1)

4 45° (1)
5 $x = 125°$ (1)
6 (a) ∠D (EDC), obtuse (1)
(b) ∠C (BCD) or ∠A (BAE), acute (1)
(c) ∠B (ABC), reflex (1)
(d) ∠E (AED), right (1)
7 (a) $x = 70°$ (1) (b) $y = 150°$ (1)
8 (a) 50° *angle drawn*, acute angle (2)
(b) 155° *angle drawn*, obtuse angle (2)
(c) 220° *angle drawn*, reflex angle (2)

R: Triangles: types and construction

1 (a) (3)

(b) angle ACB 78° (1)
2 (a) (3) (b) AC = 5.8 cm (1)

3 (a) (3)

(b) angle ACB 75° (1)
4 (a) and (b) (6)

(c) 104° (1)

5 (a) 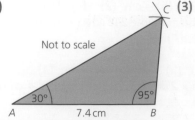 (3)   (b) 9 cm (1)

Not to scale

S: Angle facts and calculating angles

1 (a) **E:** kite (1)   (b) **C:** obtuse (1)   (c) **B:** 112.5° (1)
2 (a) $a = 40°$, $b = 140°$, $c = 140°$ (3)
   (b) $x = 110°$, $w = 70°$, $z = 110°$, $v = 70°$, $y = 110°$ (4)
   (c) $a = 75°$, $b = 48°$ (2)
3 (a) $a = 130°$ (1)   (b) $d = 66°$, $c = 66°$ (2)   (c) $e = 70°$ (1)
4 (a) $a = 45°$, angles at a point add up to 360°   (2)
   (b) $b = c = d = 60°$ (equilateral triangle), angles in a
       triangle add up to 180°   (3)
   (c) $e = f = 45°$, angles in a triangle add up to 180°
       $g = 135°$, angles on a straight line add up to
       180° $h = 22.5°$, (equilateral triangle), angles
       in a triangle add up to 180°   (6)

T: The eight-point compass

1 (a) east 090° (2)   (d) north-west 315° (2)
   (b) north-west 315° (2)   (e) north 000° (2)
   (c) north-east 045° (2)
2 (a)   (4)

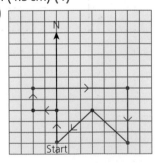

   (b) south-east   (1)
3 (a) 7 m (3.5 cm) (1)   (d) 4 m (2 cm) (1)
   (b) 6 m (3 cm) (1)   (e) 7 m (3.5 cm) (1)
   (c) 3 m (1.5 cm) (1)
4 (a)–(g)   (7)

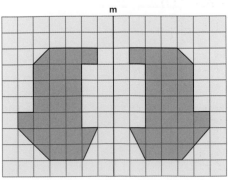

U: Reflection

1   (3)

2 (a) and (b)   (4)

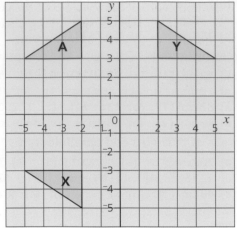

3 (a) See grid   (3)
   (b) X: (1, ⁻1), (1, ⁻3), (2, ⁻4), (4, ⁻4), (5, ⁻3), (5, ⁻1) (4)
   (c) Y: (⁻1, 1), (⁻5, 1), (⁻5, 3), (⁻4, 4), (⁻2, 4), (⁻1, 3) (4)

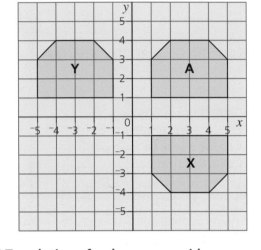

V: Translation of a shape on a grid

1 (a) 2 squares right and 2 squares down   (2)
   (b) 6 squares down   (2)
   (c) 9 squares right and 1 square up   (2)
   (d) 1 square left and 3 squares up   (2)
   (e) 3 squares left and 8 squares down   (2)
   (f) 7 squares right and 3 square up   (2)
2 (a) (i) reflection   (1)
       (ii) translation   (1)
   (b) congruent   (1)
3 (a), (b), (c)   (6)

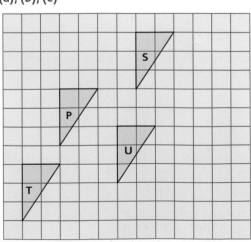

W: Rotation of a shape on a grid

1 (a) rotation of 90° clockwise about the origin   (1)
   (b) rotation of 180° clockwise about the origin   (1)
   (c) rotation of 90° anticlockwise about
       the point (⁻1, ⁻1)   (1)
   (d) rotation of 90° anticlockwise about the
       point (⁻4, ⁻3)   (1)
2 (a), (b), (c)   (6)

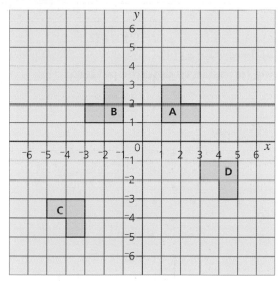

3 (a), (b), (c), (d)   (8)

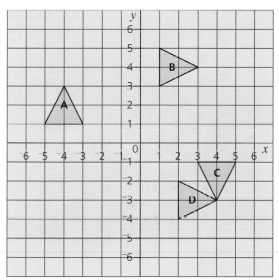

Test 4

1 (a) 2 m (1)   (b) 250 ml (1)
2 $2\frac{3}{4}$ (1)
3 6 mph   (1)
4 (a) None (1)   (b) B and D (1)
5   (1)

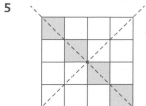

6 (a) (i) 45 mm (ii) 74 mm   (1)
   (b) (i) 68° (ii) 68° (iii) 112°   (1)
7 3 cm   (1)

8 360°   (1)
9 (a) 44 cm (1)   (b) 84 cm² (1)
10 5 cm   (1)
11 240 cm³   (1)
12 5 cm   (1)
13   (2)

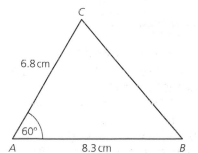

14 a = 20°, angle sum of a triangle is 180° (1)
    b = 160°, angles on a straight line add up to 180° (1)
    c = 90°, angles on a straight line add up to 180° (1)
15 (a), (b), (c), (d)   (4)

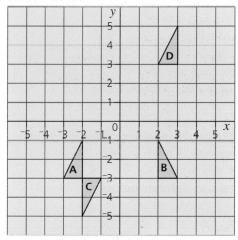

5 Algebra

A: Missing numbers; missing operations

1 (a)   (1) (c)   (1)

1	3	8	0
+		1	5	6
	1	5	3	6

	1	5	8
×			2
	3	1	6

   (b)   (1)

7	5	8
−		4	2
	7	1	6

2 (a) 34 − 4 = 30 (1)   (c) 12 ÷ 6 × 4 = 8 (1)
   (b) 19 = 26 − 7 (1)
3 (a)   (1)

1	0	6	2
×			5
	5	3	1	0

   (b)   (1)

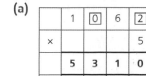

(c) 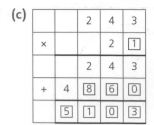 (1)

		2	4	3
×			2	[1]
		2	4	3
+	4	[8]	[6]	[0]
	[5]	[1]	[0]	[3]

4 (a) $30 \div 3 = 10$ (1)
   (b) $4 \times 3 = 8 + 4$ (1)
   (c) $(12 \times 2) - 6 = 9 \times 2$ (1)
5 (a) $8 + 7 - 4 = 11$ (1)
   (b) (1)

		[2]	[5]	6
−		1	5	[4]
		1	**0**	**2**

6 (a) ✚ = 3 (1), ◆ =15 (1)
   (b) ● = 5, ◆ = 4 (3)
7 6, 2 (2)
8 £6 (2)

B: Function machines

1 (a) $a = 7, b = 5, c = 1$ (3)
   (b) $a = ^-6, b = 5, c = 10$ (3)
2 (a) $a = 3, b = 10, c = 6, d = 12, e = ^-1$ (5)
   (b) $a = 10, b = 2, c = -10, d = ^-2, e = 20, f = 8$ (6)
3 (a) $a = ^-4, b = ^-11, c = 0$ (3)
   (b) $a = 12, b = 3, c = ^-4, d = ^-1, e = 38, f = 10$ (6)
4 (a) $A = -3, a = ^-3, b = 6$ (3)
   (b) $B = + 4, C = \div 6, a = 3, b = ^-10$ (4)

C: Sequences and number puzzles

1 (a) (Rule: + 5) 26, 31, 36 (1)
   (b) (Rule: − 7) 12, 5, ⁻2 (2)
   (c) (Rule: + 0.2) 5.0, 5.2, 5.4 (2)
2 (a) (Rule: + 8) 51 59 (1)
   (b) (Rule: − 7) 65 58 (1)
   (c) (Rule: − 3) 0 ⁻3 (2)
3 (a) (Rule: + 5) 5, 10 (1)
   (b) (Rule: + 1, + 2, + 3, …) **9**, 10, 12, 15, 19, **24**, 30 (1)
4 (a) (Rule: add a sequence of odd numbers) 27, 38 (2)
   (b) (Rule: divide by 2) 10, 5 (1)
   (c) (Rule: $n \times 2$)+ 1 47, 95 (2)
5 (a) (Rule: + 0.25) 0.5, **0.75**, 1, 1.25, **1.5**, 1.75 (2)
   (b) (Rule: sequence of square numbers) 16, **25**, 36,
     49, 64, 81 (2)
6 4 and 6 (6 + 4 = 10, 6 × 4 = 24) (2)
7 4 and 15 (15 − 4 = 11, 15 × 4 = 60) (2)

D: Formulae

1 12 (1)
2 (a) $x - 5$ (1) (b) $10y$ (1)
3 (a) 100 (1)
   (b) (i) 500 cm (1) (ii) 900 cm (1)
   (c) $100n$, where n is the number of metres (1)
4 (a) $3(a + 4)$ (1) (b) 30 [$3(a + 4) = 3 \times 10$] (1)
5 (a) (i) 15 minutes
     (ii) 2 hours or 120 minutes (2)
   (b) $T = \frac{n}{40}$ (2)
6 (a) $(a \times 7) + 6$ (1) (b) 41 [$(5 \times 7) + 6$] (1)
7 (a) $\frac{b - a}{2}$ (1) (b) $\frac{c}{d} + e$ (1)
8 (a) $2n + 5$ (1) (b) 1 (1)

E: Equations

1 (a) $M = 11 + 2$, Maya is 13 years old (1)
   (b) $6a = 42$, side is 7 cm (1)
2 (a) ■ = 16 (1) (b) ◆ = 15 (1) (c) ✱ = 4 (1)
3 (a) $a = 7$ (1) (b) $b = 12$ (1) (c) $c = 5$ (1)
4 (a) ■ = 4 (1) (b) ✱ = 9 (2)
5 (a) $d = 2$ (2) (b) $e = 7$ (3)
6 (a) $v = 14$ (1) (c) $x = 12$ (3)
   (b) $w = 24$ (2)
7 1 (1)
8 (a) $l + a + i$ (1) (c) 17 (2)
   (b) 14 (2 + 5 + 7) (1)

F: Using formulae to describe patterns

1 (2)

4	5

2 (a) (3)

Pattern number	Number of dots
1	3
2	6
3	9
4	12
5	15

   (b) Total number of dots in nth pattern = $3n$ (1)
3 (a) (2)

4	5

   (b) (6)

Pattern number	Green squares	White squares	Total squares
1	2	2	4
2	4	4	8
3	6	6	12
4	8	8	16
5	10	10	20

   (c) Total number of squares in the nth pattern = $4n$ (2)
   (d) Green squares in the nth pattern = $2n$ (2)
   (e) White squares in the nth pattern = $2n$ (2)

Test 5

1 (a) 13 (1) (b) 0 (1)
2 (a) 4 (1) (b) 1 (1)
3 (a) (i) + 3 (ii) 18, 21 (1)
   (b) (i) × 2 (ii) 32, 64 (1)

    (c) (i) − 4 (ii) 17, 13   (1)
    (d) (i) − 4 (ii) 2, ⁻2   (1)

4 (a) 22, 29 (1) (b) 29, 47 (1)

5 30, 2   (2)

6 15   (1)

7 14   (1)

8 (a) $\frac{6}{d}$ (1) (b) $h(g - f)$ (1)

9 $x = 9$ (2)

10 (a)   (2)

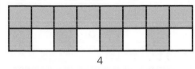

4

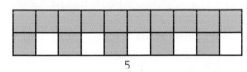

5

(b)   (3)

Pattern number	Pink squares	White squares	Total squares
1	3	1	4
2	6	2	8
3	9	3	12
4	12	4	16
5	15	5	20

(c) Total squares in the nth term = $4n$ (1)
(d) Total pink squares in the nth term = $3n$ (1)

6 Data handling

A: Carroll diagrams and Venn diagrams

1 (a) 4   (2)
    (b) 6   (2)
    (c) 18   (3)

2 (a)   (4)

	divisible by 3	not divisible by 3
odd	3, 9, 15	1, 5, 7, 11, 13, 17, 19
not odd	6, 12, 18	2, 4, 8, 10, 14, 16, 20

(b)   (4)

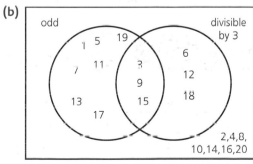

3 (a)   (4)

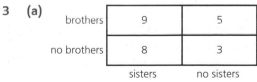

(b)   (4)

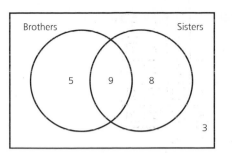

4 (a) 2 × 2 × 2 × 2   (2)
    (b) 2 × 2 × 2 × 3   (2)
    (c)   (2)

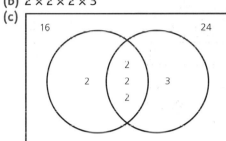

(d) 2, 2, 2 (1) (e) 8 (1)

B: Pictograms

1 (a) Oriel (1) (c) 5 marks (1)
    (b) Maya (1) (d) 2 pupils (1)

2 (a) Dorothy (1)
    (b) £20 (Dorothy: £30 and Toby: £10) (1)
    (c) £100 (1) (d) Dorothy and Esther (1)

3 (a) 24 bottles (1) (d) 90 bottles (2)
    (b) 18 bottles (1)
    (c) *Check chart shows 5 full circles on Friday* (2)

C: Bar charts

1 (a) 14 dogs (1) (c) 17 (1)
    (b) 19 cats (1) (d) 41 in total (2)

2 (a) 175 g (2) (b) 0.125 kg (2) (c) 175 g (2)
    (d)   (2)

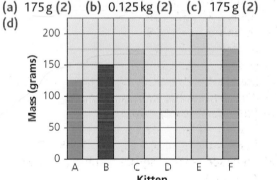

3 (a) January (1) (b) August (1)
    (c) (i) April and May   (1)
        (ii) 1200 ice creams   (2)
    (d) Ice creams are bought more often in the hotter summer months.   (1)

D: Frequency diagrams

1 (a)   (4)

Goals scored in a match	Tally	Frequency
0	III	3
1	III	3
2	ℍI	5
3	II	2
4	I	1
5	I	1
	Total	15

(b) ■ Goals scored in a match (3)

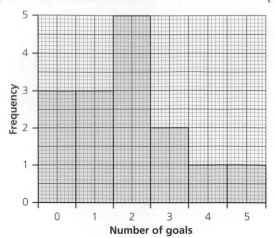

2 **(a)** 3 (1)
   (b) 10 (2)
   (c) 75 (1)
   (d) 85 (1)

3 **(a)** (5)

Result	Tally	Frequency
2	I	1
3	II	2
4	II	2
5	II	2
6	II	2
7	IIII	4
8	III	3
9	II	2
10	II	2
	Total	20

(b) ■ Test results (5)

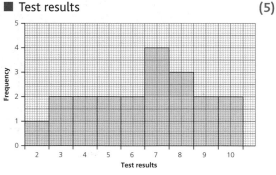

(c) 20 (1)   **(d)** 13 (1)

E: Pie charts

1 **(a)** $\frac{1}{4}$ (1)   **(b)** 25% (1)   **(c)** 75% (1)   **(d)** 50 (2)

2 **(a)** 25% (1)   **(b)** $\frac{1}{8}$ (1)   **(c)** $\frac{3}{8}$ (2)   **(d)** 48 (2)

3 **(a)** $\frac{1}{4}$ (1)   **(b)** $\frac{1}{3}$ (1)   **(c)** 70° (1)   **(d)** $\frac{7}{36}$ (1)

   (e) **(i)** 18 beetles (2)
       (ii) 24 earwigs (2)
       (iii) 4 ladybirds (2)

F: Drawing pie charts

1 (10)

Fruit	Frequency	Fraction	Angle
Banana	5	$\frac{5}{30} = \frac{1}{6}$	60°
Apple	9	$\frac{9}{30} = \frac{3}{10}$	108°
Satsuma	6	$\frac{6}{30} = \frac{1}{5}$	72°
Grapes	10	$\frac{10}{30} = \frac{1}{3}$	120°
Total	30		360°

2 ■ Fruit in lunch boxes (5)

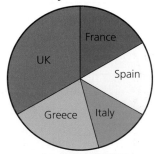

3 **(a)** (10)

Destination	Frequency	Fraction	Angle
United Kingdom	8	$\frac{8}{24} = \frac{1}{3}$	120°
France	4	$\frac{4}{24} = \frac{1}{6}$	60°
Italy	3	$\frac{3}{24} = \frac{1}{8}$	45°
Spain	4	$\frac{4}{24} = \frac{1}{6}$	60°
Greece	5	$\frac{5}{24}$	75°
Total	24		360°

(b) ■ Holiday destinations (5)

(c) Italy (1)   **(d)** United Kingdom (1)

G: Range, mode, median and mean

1 **(a)** 1 2 3 4 4 4 5 5 6 6 (1)
       (i) 5 (6 − 1) (1)   **(ii)** 4 (1)   **(iii)** 4 (1)
       (iv) 4 (40 ÷ 10) (2)
   (b) 23 25 29 34 34 43 45 55 (1)
       (i) 32 (1)   **(ii)** 34 (1)   **(iii)** 34 (1)
       (iv) 36 (288 ÷ 8) (2)

2 **(a)** 7 (1)   **(b)** 7 (1)   **(c)** 6 (1)   **(d)** 6 (96 ÷ 16) (2)
3 12 °C (2)

11 + Mathematics Revision Guide published by Galore Park

4 (a) (5)

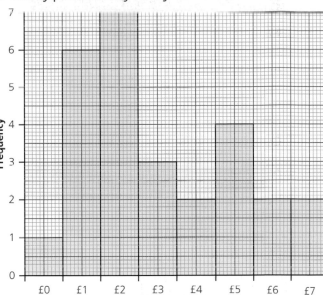

Weekly pocket money survey

(b) 25 (1) **(d)** £2 (1)   **(f)** £2.80 (70 ÷ 25) (2)
(c) £7 (1) **(e)** £2 (2)

H: Grouped data

1 **(a)** 23 (78 − 55) (2)
  (b) 66 (half way between 65 and 67, the midpoints
  of the ordered list) (3)
  (c) (4)

Waist measurement (cm)	Tally marks	Frequency
55–57	II	2
58–60		0
61–63	IIII I	6
64–66	II	2
67–69	IIII	5
70–72	III	3
73–75	I	1
76–78	I	1
Total		20

2 **(a)** (4)

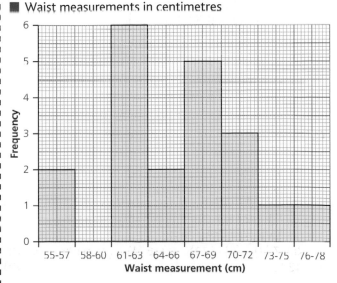

Waist measurements in centimetres

(b) 61–63 cm (1)

3 **(a)** (5)

Number of goals scored in each game

(b) 10–12 goals (1)
(c) 8 goals (1)   **(d)** 15 games (1)

I: Line graphs

1 **(a)** 13:00 (1)   **(c)** 15:00 (1)
  (b) 20 litres (1)   **(d)** 50 litres (2)
2 **(a)** 19°C (1)   **(c)** 9 (1)
  (b) 15:00 (1)   **(d)** 23.5°C (3)
3 **(a)** 2.6°C (1)   **(b)** 07:00 (1)   **(c)** 9 (2)

J: Drawing line graphs

1 **(a), (b), (c)** (7)

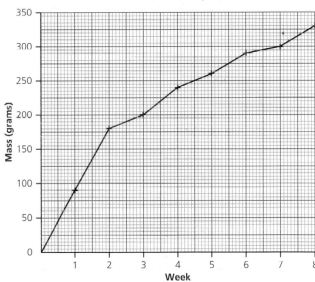

Masses of kittens over an 8-week period

(d) between weeks 1 and 2 (1)

2 (a), (b), (c)   (7)

■ Distance travelled to school

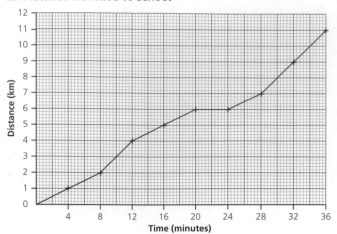

(d) The car stopped for 4 minutes, perhaps at traffic lights.   (1)

3 (a), (b), (c)   (7)

■ Water temperature during heating

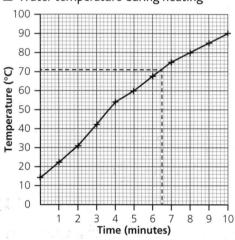

(d) 71°C   (1)

K: Conversion graphs

1 (a) £12 (2)   (c) 10 metres (2)
    (b) £3.75 (2)   (d) 3.3 metres (2)

2 (a) £0.20 (1)   (c) $8 (1)
    (b) $0.20 (1)   (d) £5.60 (1)

3 (a) 0 km (1)   (b) 16 km (1)   (c) 80 km (1)

    (d)   (5)

■ Conversion of miles and kilometres

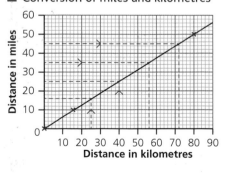

(e) (i) 25 miles (2)   (iii) 56 km (2)
    (ii) 16 miles (2)   (iv) 72 km (2)

L: Probability

1 (a) 1, 2, 3, 4, 5, 6   (2)

    (b) (i) $\frac{1}{2}$ (1)   (ii) $\frac{1}{6}$ (1)   (c) 50% (1)

2 (a) $\frac{4}{8} = \frac{1}{2}$ (1)   (b) $\frac{2}{8} = \frac{1}{4}$ (1)   (c) $\frac{3}{8}$ (2)

    (d) $\frac{4}{8} = \frac{1}{2}$ (E and O have reflection symmetry)   (1)

3 (a) 2 (Heads and Tails)   (1)
    (b) 5 (A, B, C, D, E)   (1)

    (c)   (5)

		Spinner				
		A	B	C	D	E
Coin	Heads (H)	HA	HB	HC	HD	HE
	Tails (T)	TA	TB	TC	TD	TE

    (d) 10 (1)   (e) $\frac{1}{10}$ (1)   (f) $\frac{5}{10} = \frac{1}{2}$ (1)

Test 6

1 (a)   (1)

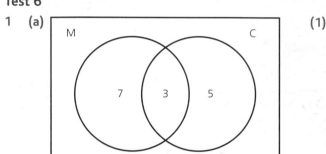

    (b) 3   (1)

2 (a) 1 girl (1)   (c) Year 4 (1)   (e) 6 (1)
    (b) 4 boys (1)   (d) 44 (1)

3 (a) 7 (1)   (b) 5 (1)   (c) 6 (1)   (d) 5 (1)

    (e)   (2)

■ Number of books read by six children

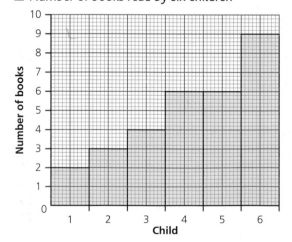

4 (a) 35   (1)
    (b) 7 years   (1)
    (c) 16   (1)
    (d) 10 years   (1)

11 + Mathematics Revision Guide published by Galore Park

(e)   (2)

■ Ages of children in a cricket club

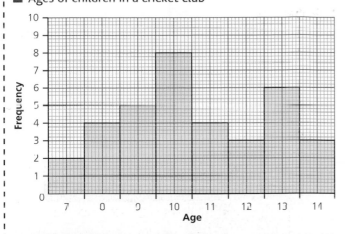

5   (a)   36 possible outcomes in total   (2)

		Second die					
		1	2	3	4	5	6
First die	1	1, 1	1, 2	1, 3	1, 4	1, 5	1, 6
	2	2, 1	2, 2	2, 3	2, 4	2, 5	2, 6
	3	3, 1	3, 2	3, 3	3, 4	3, 5	3, 6
	4	4, 1	4, 2	4, 3	4, 4	4, 5	4, 6
	5	5, 1	5, 2	5, 3	5, 4	5, 5	5, 6
	6	6, 1	6, 2	6, 3	6, 4	6, 5	6, 6

   (b) (i)   3, 1   2, 2   1, 3 (shown in green)   (1)

      (ii) $\frac{3}{36} = \frac{1}{12}$ (1)

   (c) (i)   6, 2   5, 3   4, 4   3, 5   2, 6

          (shown in pink)   (1)

      (ii) $\frac{5}{36}$ (1)

Sample 11+ test

1   ⁻3   ⁻2.50   ⁻1   0.52   1.25   2.52   (1)
2   (a) 12.6 (1)   (c) 13 (1)
    (b) 12.58 (1)   (d) 13 (1)
3   (a) 36 (6 × 6) or 64 (8 × 8) (1)   (c) 27 (3 × 3 × 3) (1)
    (b) 36 (1)   (d) 9, 5 (9 × 5 = 45) (1)
4   (a) 2 (4 = 2 × 2, 12 = 6 × 2, 18 = 9 × 2)   (1)
    (b) 20 (60 = **2 × 2** × 3 × 5 and
          80 = **2 × 2** × 2 × 2 × **5**)   (1)
5   £1 (50p × 2 = 100p and 20p × 5 = 100p)   (1)
6   21 (26 − 18 + 13 = 21)   (1)
7   ⁻8   (1)
8   34 [BIDMAS: 12 + (11 × 2) = 12 + 22]   (1)
9   £269 (£84 + £110 + £75)   (3)
10   (1)

TH	H	T	U
⁴5̷ ¹³4̷	¹5̷	¹3	8
−	6	5	2
4	7	8	6

11   (1)

	TTH	TH	H	T	U	
			3	4	5	
×				1	4	
		1	3	8	0	× 4
+		3¹	4¹	5²	0	×10
		4	8	3	0	
			1			

12   8 (factors of 32 are 2 × 2 × 2 × 2 × 2, so divide 256 by 2 five times)   (1)
13   125 r 12 or 125.5   (1)

		Th	H	T	U	
		0	1	2	5	
2	4	3	0	1	2	
		2	4			
			6	1		
			4	8		
			1	3	2	
			1	2	0	
				1	2	

14   £3172.50 (£6345 ÷ 2)   (1)
15   D (42 × 8 = 336)   (2)
16   (a) $\frac{2}{3}$ (1)   (b) $\frac{1}{3}$ (1)   (c) $\frac{1}{2}$ (1)
17   (a) $1\frac{1}{2}$ (1)   (b) $4\frac{2}{3}$ (1)   (c) $7\frac{2}{5}$ (1)
18   $\frac{3}{4}$ $(\frac{12}{16})$ (1)
19   $3\frac{1}{4}$ $(\frac{5}{12} + 2\frac{5}{6} = \frac{5}{12} + \frac{17}{6} = \frac{5}{12} + \frac{34}{12} = \frac{39}{12} = 3\frac{3}{12})$ (1)
20   $2\frac{1}{2}, 3\frac{1}{4}$ (1)
21   1:3   (1)
22   (a) 3:7   (1)
     (b) 7:10   (1)
23   (a) 10 miles (1 km = $\frac{5}{8}$ miles, so 16 × $\frac{5}{8}$ = 10)   (1)
     (b) 25 km   (1)
24   (a) 1800 cm or 18 m (36 × 50)   (1)
     (b) 10.5 cm (525 ÷ 50)   (1)
25   (a) £3450   (1)
     (b) 0.2345   (1)
26   9 months (£8 × 8 months = 64, so he needs to save another 1 month to get enough)   (1)
27   (a) $2\frac{1}{4}$ (1)

     (b) $14\frac{3}{1000}$ (1)

     (c) $\frac{1}{500}$ $(\frac{2}{1000})$ (1)
28   C: 5 $(\frac{3}{4}, \frac{15}{20}, \frac{75}{100}, \frac{27}{36}, \frac{9}{12})$ (4)

29 (5)

Fraction (in simplest form)	$\frac{2}{5}$	$1\frac{1}{4}$	$\frac{4}{25}$	$\frac{3}{5}$	$2\frac{1}{4}$
Decimal	0.4	1.25	0.16	0.6	2.25
Percentage	40%	125%	16%	60%	225%

30 $\frac{1}{4}$ 22% 0.208 $\frac{1}{5}$ (2)

31 £86 (1)

32 £1452 (£132 × 11) (1)

33 14 g $(10 \times \frac{7}{5})$ (1)

34 18 kg (1)

35 £80 (20% of £100 = £20, £100 − £20) (1)

36 £21 (16% of £25 = £4, £25 − £4) (1)

37 (a) $\frac{1}{20}$ (1)   (b) $1\frac{1}{8}$ (1)

38 (a) 15 $(6 \times \frac{5}{2})$ (1)

(b) $2\frac{3}{8} (\frac{3}{4} \times \frac{19}{6})$ (1)

39 $\frac{1}{9}$ litre $(\frac{2}{3} \div 6 = \frac{2}{3} \times \frac{1}{6} = \frac{2}{18})$ (1)

40 $\frac{9}{10}$ cups of flour $(\frac{15}{25} \times \frac{3}{2})$ (1)

41 (a) (i) 50 m*l* (1)
   (ii) 25 m*l* (1)

(b) (2)

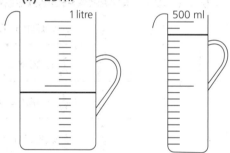

42 (a) 425 cm (1)   (b) 4250 mm (1)

43 1 hour and 45 minutes (1)

44 35 mph (1)

45 30 km (1)

46 B, C and E (1)

47 (a) Quadrilaterals (1)   (c) 4 (1)   (e) Right angles (1)
   (b) Rhombus (1)   (d) None (1)   (f) 4 (1)

48 (1)

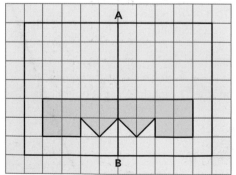

49 (a) (i) 90° (1)
   (ii) 90° (1)
   (iii) 135° (1)

(b) *CD* = **2 cm** and *AD* = **5.6 cm** *(Drawn at 50%)* (4)

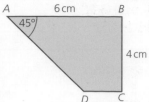

50 (a) 6   (b) order 6 (2)

51 9 cm (1)

52 (a) 300 m (1)   (b) £750 (1)
   (c) 3800 m² (60 × 50 = 3000, 40 × 20 = 800, 3000 + 800) (1)

53 18 cm² (1)

54 22 m (1)

55 23 cm (1)

56 A and B (1)

57 24 m³ (1)

58 6 m (1)

59 (a) x = 120° (1)   (b) y = 175° (1)

60 (a) (2)

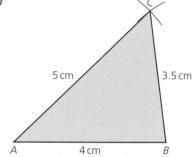

(b) 53° (1)

61 a = 45°, b = 45°, c = 135°, d = 22.5° (1)

62 (a)–(g) (5)

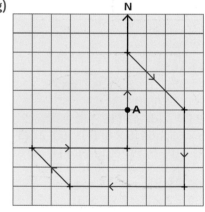

63 (a), (b),   (⁻1, 2), (⁻3, 2), (⁻2, 4)
   (c), (d) (5)

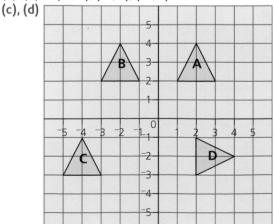

64 (a) ■ = 7 (1) (c) ■ = 8 (1)
   (b) ● = 52 (1) (d) ☆ = 13 (1)
65 (a) 20 × 3 = 60 (b) 5 × 3 = 20 − 5   (2)
66 £2.40   (1)
67 $a = 3, b = {}^-10, c = {}^-1$ (3)
68 A = − 2, B = × 3, a = 36, b = ⁻6   (4)
69 52   46   (1)
70 ⁻0.5, ⁻0.25, 0, 0.25, 0.5, 0.75   (1)
71 12 and 3   (1)
72 (a) $12a + 4$ (1) (b) 40 (1)
73 18   (1)
74 $a = 14$ (1)
75 3   (1)
76 (a)   (2)

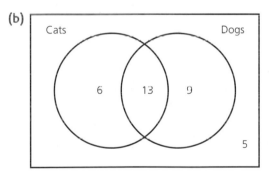

4   5

(b)   (3)

Pattern number	Purple squares	Green squares	Total squares
1	1	3	4
2	2	6	8
3	3	9	12
4	4	12	16
5	5	15	20
6	6	18	24

(c) $3n$ (1)

77 (a)   (2)

	cats	no cats
dogs	13	9
no dogs	6	5

(b)   (2)

Cats   Dogs

6   13   9

5

78 (a) Archie (1) (b) 15 (1) (c) 2 (1)

79 (a)   (3)

■ Numbers of parents watching rugby matches

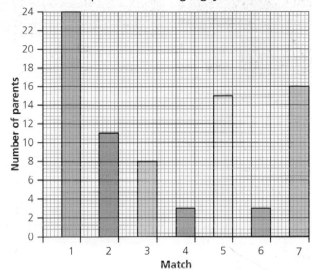

(b) 21 (24 − 3)   (1)
(c) 11 (3 3 8 **11** 15 16 24)   (1)

80
(a)   (3)

Goals scored in a match	Tally	Frequency	Fraction	Angle
0	III	3	$\frac{1}{5}$	$\frac{1}{5} \times 360° = 72°$
1	I	1	$\frac{1}{15}$	$\frac{1}{15} \times 360° = 24°$
2	I	1	$\frac{1}{15}$	$\frac{1}{15} \times 360° = 24°$
3	III	3	$\frac{1}{5}$	$\frac{1}{5} \times 360° = 72°$
4	IIII	5	$\frac{1}{3}$	$\frac{1}{3} \times 360° = 120°$
5	II	2	$\frac{2}{15}$	$\frac{2}{15} \times 360° = 48°$
	Total	15		360°

(b) ■ Goals scored in hockey matches   (2)

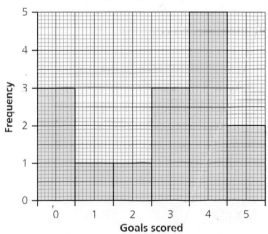

(c)

0 = 72°   (2)
1 = 24°
2 = 24°
3 = 72°
4 = 120°
5 = 48°

■ Goals scored in hockey matches

81 (a) 1, 2, 3, 4, 5, 6   (1)

   (b) (i) $\frac{1}{2}$ (1)   (ii) $\frac{1}{3}$ (1)

   (c) 50 times   (1)

82 (a)   (1)

	Die (1)	Die (2)	Die (3)	Die (4)	Die (5)	Die (6)
Coin – Heads (H)	H1	H2	H3	H4	H5	H6
Coin – Tails (T)	T1	T2	T3	T4	T5	T6

   (b) 12   (1)

   (c) $\frac{3}{12} = \frac{1}{4}$ (shown in blue)   (1)

   (d) $\frac{6}{12} = \frac{1}{2}$ (shown in green)   (1)

83 (a) 9 10 12 13 14 14 14 15 15 16 17 19   (1)
   (b) 10 (1)   (c) 14 (1)   (d) 14 (1)   (e) 12 (1)

84 (a), (b)   (2)

■ Water temperature

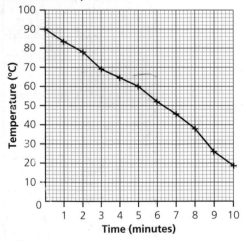

   (c) 49°   (1)

85 (a) 2 cents (1)   (b) £5 (1)   (c) $10.80 (1)